Student Support Materials for AQA

AS/A-level Year 1

Physics

Sections 4 and 5: Mechanics and materials, Electricity

Author: Dave Kelly

William Collins' dream of knowledge for all began with the publication of his first book in 1819.

A self-educated mill worker, he not only enriched millions of lives, but also founded a flourishing publishing house. Today, staying true to this spirit, Collins books are packed with inspiration, innovation and practical expertise. They place you at the centre of a world of possibility and give you exactly what you need to explore it.

Collins. Freedom to teach

HarperCollins Publishers
The News Building
1 London Bridge Street
London SE1 9GF

Browse the complete Collins catalogue at www.collins.co.uk

10 9 8 7 6 5 4 3 2 1

ISBN 978-0-00-818952-5

www.collins.co.uk

A catalogue record for this book is available from the British Library

Commissioned by Gillian Lindsey
Edited by Alexander Rutherford
Project managed by Maheswari PonSaravanan at Jouve
Development by Aidan Gill
Copyedited and proof read by Janette Schubert
Typeset by Jouve India Private Limited
Original design by Hedgehog Publishing
Cover design by Angela English
Production by Lauren Crisp
Printed by CPI Group (UK) Ltd, Croydon, CR0 4YY
Cover image © Shutterstock/WHITE RABBIT83

Contents

3.4	**Mechanics and materials**	**4**
	3.4.1 Force, energy and momentum	4
	3.4.2 Materials	39
3.5	**Electricity**	**51**
	3.5.1 Current electricity	51
	Examination preparation	
	Practical and mathematical skills	74
	Data and formulae	76
	Practice exam-style questions	79
	Answers	89
	Glossary	**94**
	Index	**99**
	Notes	**101**

3.4 Mechanics and materials

3.4.1 Force, energy and momentum

3.4.1.1 Scalars and vectors

Physical quantities can be classified into two groups: **scalars** or **vectors**. Scalar quantities, such as temperature or mass, have a magnitude (size) but have no direction associated with them. Scalars can be fully described by a single number and a unit. For example, stating that the room temperature is 20°C or that the mass of a person is 80 kg fully specifies these quantities. Other physical quantities, like velocity or force, have a direction associated with them. These are known as vector quantities. A vector quantity is only fully specified when the magnitude *and* the direction are given. It isn't sufficient to know that a force has a magnitude of 300 N; we also need to state what direction it acts in, e.g. a force of 300 N acting horizontally in a direction 30° east of north.

Notes

If you are asked to find an unknown vector, such as force, don't forget to give the direction as well as the magnitude.

Table 1
Examples of scalar and vector quantities

Scalars	Vectors
distance	displacement
speed	velocity
energy	force
power	acceleration
mass	momentum

Notes

Make sure that you know which quantities are vectors and which are scalars.

Definition

A vector quantity has magnitude and direction, whereas a scalar quantity has magnitude only.

Vector quantities are often identified by the use of **bold type**.

Adding vector quantities

When two vectors are added, we need to take account of their direction as well as their magnitude. Two vectors can be added by drawing a scale diagram showing one vector followed by the other, i.e. by drawing them 'nose to tail' (see Fig 1). It does not matter which vector you draw first.

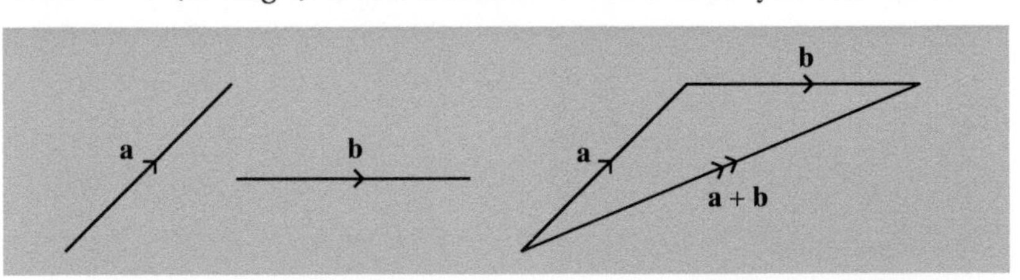

Fig 1
Adding vectors
The sum of **a** and **b** is found by drawing **a** and **b** so that the arrows showing their direction follow on.

The sum of a number of vectors is known as the **resultant**. The resultant is the single vector that has the same effect as the combination of the other vectors. It is vital to take into account the relative direction of vectors when adding them together, for example the resultant of two 5 N forces could be anything from zero to 10 N, depending on their directions (see Fig 2).

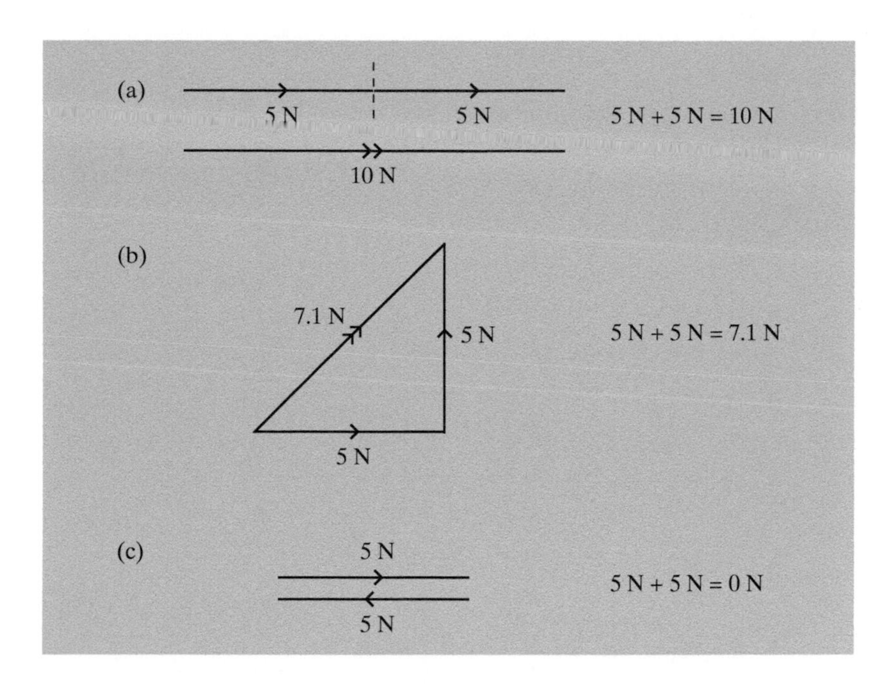

Fig 2
The magnitude and direction of the resultant depends on the magnitude and direction of the two component vectors.

The resultant of two vectors can also be found by the **parallelogram law**. A parallelogram is constructed using the two vectors as adjacent sides. The resultant is the diagonal of the parallelogram (Fig 3).

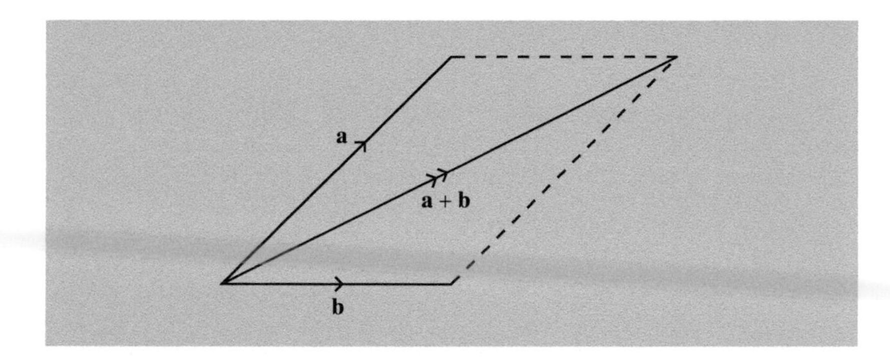

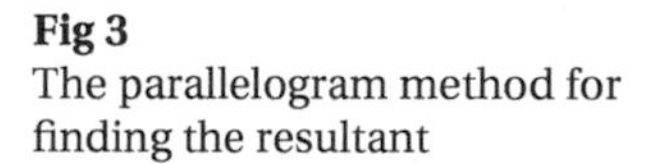
Fig 3
The parallelogram method for finding the resultant

Notes

The resultant of two vectors at any angle can be found using the sine and cosine rules, but questions at A-level are restricted to perpendicular vectors.

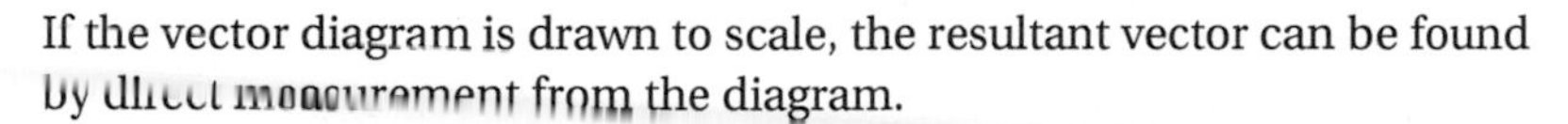
If the vector diagram is drawn to scale, the resultant vector can be found by direct measurement from the diagram.

For two vectors at right angles, the magnitude of the resultant can also be found from calculation using Pythagoras' theorem (Fig 4).

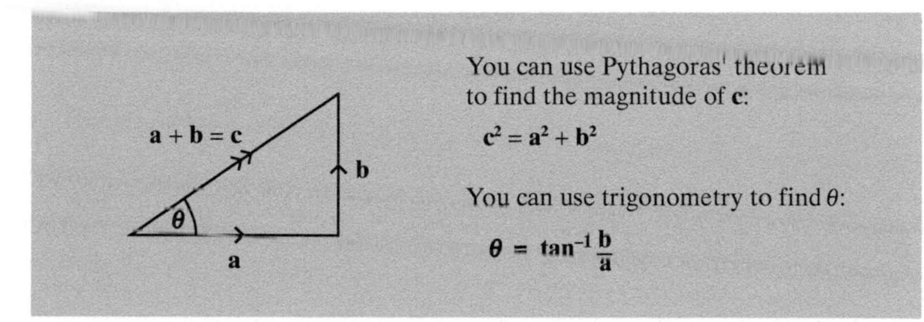

Fig 4
Adding vectors at right angles

Example

A tanker is being pulled into harbour by a tug boat which exerts a force of 200 MN in an easterly direction. The tanker is also subject to a force of 150 MN due to a northerly current. Find the resultant force acting on the tanker.

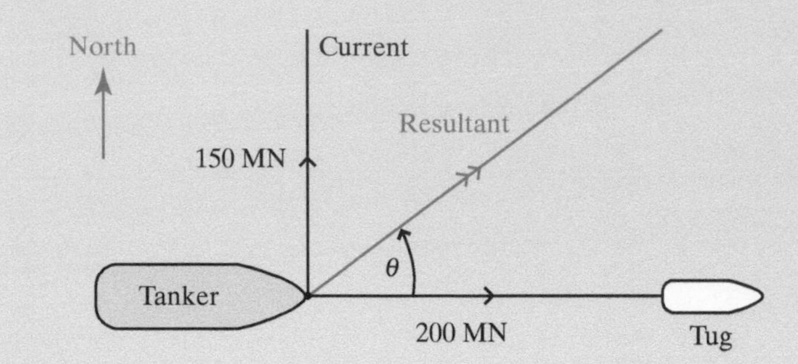

Answer

The magnitude of the resultant, R, is given by the equation:

$$R^2 = 200^2 + 150^2$$
$$= 40\,000 + 22\,500$$
$$= 62\,500$$
$$R = 250 \text{ MN}$$

Find the direction of the resultant by:

$$\theta = \tan^{-1}\frac{150}{200}$$
$$= \tan^{-1} 0.75$$
$$= 37° \text{ north of east}$$
(or 53° east of north)

Fig 5
Subtracting a vector

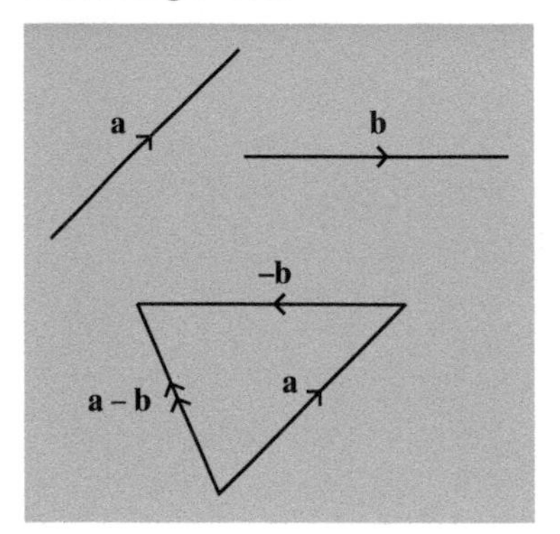

Subtracting a vector quantity can be thought of as adding a negative vector. The vector which is to be subtracted is reversed in direction. This reversed, or negative, vector is then added in the usual way (Fig 5).

Example

A river is flowing at 1.0 m s^{-1}. Find the speed and direction that a swimmer must travel if he is to achieve a resultant velocity of 1.5 m s^{-1} directly across the river.

Answer

The swimmer's velocity, v, is the resultant minus the river's velocity.

$$\text{Magnitude of velocity} = \sqrt{1.5^2 + 1.0^2}$$
$$= 1.8 \text{ m s}^{-1}$$

$$\text{Direction of velocity, } \theta = \tan^{-1}\frac{1.0}{1.5}$$
$$= 34°$$

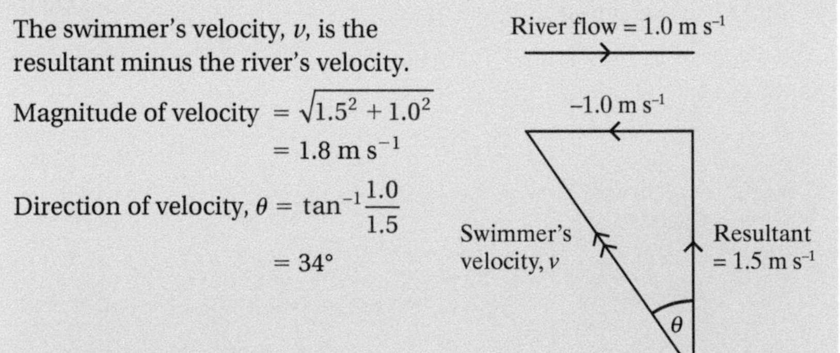

Resolution of vectors

A single vector can be replaced by a combination of two or more vectors that would have the same effect. This process is called **resolving** the vector into its **components** and it can be thought of as the reverse of finding the resultant. The components of a vector could be at any angle but it is often useful to use two components that are at right angles to each other. This might be to find the horizontal and vertical components of a force or a velocity (Fig 6).

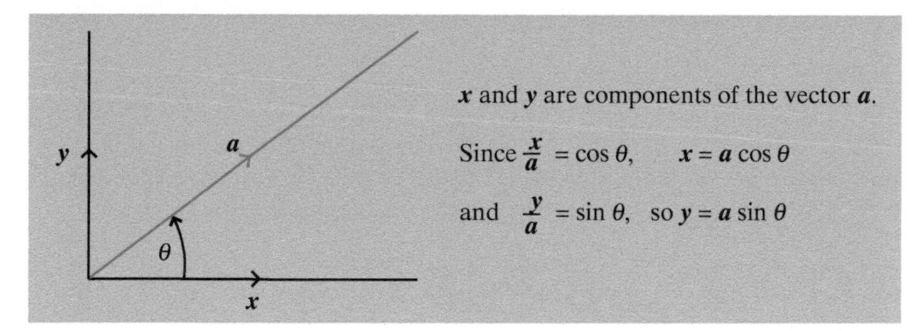

Fig 6
Calculating perpendicular components

Example

A wind is blowing at 15.0 m s^{-1} in a north-easterly direction. Find the components of the wind's velocity which blow towards the north and towards the east.

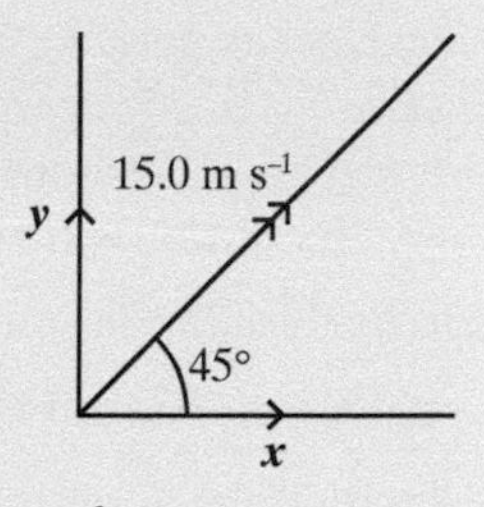

Answer

The northerly component, $y = 15.0 \sin 45° = 10.6$ m s^{-1}

The easterly component, $x = 15.0 \cos 45° = 10.6$ m s^{-1}

Example

A car of weight 10 000 N is parked on a steep hill which makes an angle of 20° to the horizontal. Resolve the car's weight into components that act along the slope and at 90° to the slope.

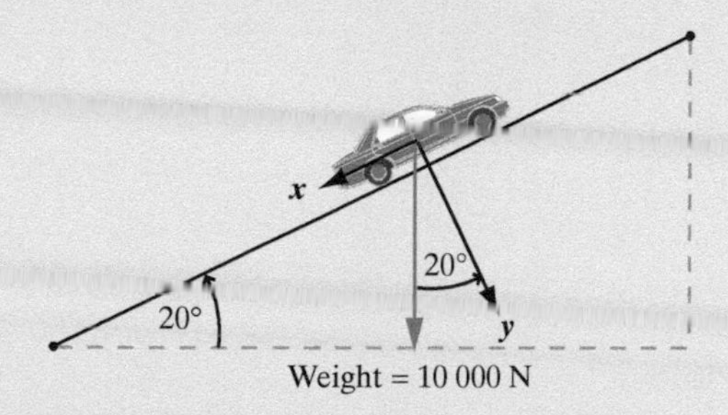

Answer

Down the slope, component $x = 10\,000 \sin 20° = 3420$ N

Perpendicular to the slope, component $y = 10\,000 \cos 20°$

$= 9397$ N $= 9400$ N (to 3 s.f.)

Essential Notes

Remember that the vector you are resolving is always the hypotenuse of a triangle. The components will always be smaller than the original vector.

Notes

In this specification, problems will be restricted to systems of two or three **coplanar forces**. Coplanar forces act in the same plane. In other words you only need to consider two-dimensional problems.

Essential Notes

Gravitational field strength varies slightly from place to place on the Earth's surface, and decreases as you move away from the Earth.

Gravitational field strength on the Moon is about 1.6 N kg^{-1}, about one sixth of its value on Earth. On the Moon, you would weigh one sixth of your weight on Earth, though your mass would remain the same.

Gravitational field strength, measured in N kg^{-1}, is represented by the letter g. The symbol g is also used to represent the acceleration due to gravity (page 21) and can be given in m s^{-2}. These quantities are numerically equivalent.

Analysing the forces

It is often important to be able to identify, and add together, all the forces that are acting on an object. The size and direction of the resultant force will determine what happens to the object.

Everyday objects are subjected to a variety of forces, such as weight, **contact forces, friction**, tension, **air resistance** and buoyancy. All these forces, except weight, are electromagnetic in origin. They arise because of the attraction or repulsion of the charges in atoms.

Weight

This is the force that acts on a mass due to the gravitational attraction of the Earth. The gravitational field strength, g, on Earth is 9.81 N kg^{-1}. This means that each kilogram of mass is attracted towards the Earth with a force of 9.81 newtons. The weight of an object (in **newtons**) is given by:

weight (N) = mass (kg) × gravitational field strength (N kg^{-1}) *or* $W = mg$

Contact forces

Whenever two solid surfaces touch, they exert a contact force on each other. This force is often known as the reaction. The contact force between the floor and your feet stops gravity pulling you through the floor. The resultant contact force between two surfaces can be at any angle (Fig 7).

We usually split the contact force into two components:

- the normal contact force acting perpendicularly to the two surfaces
- the frictional force, acting parallel to the surfaces.

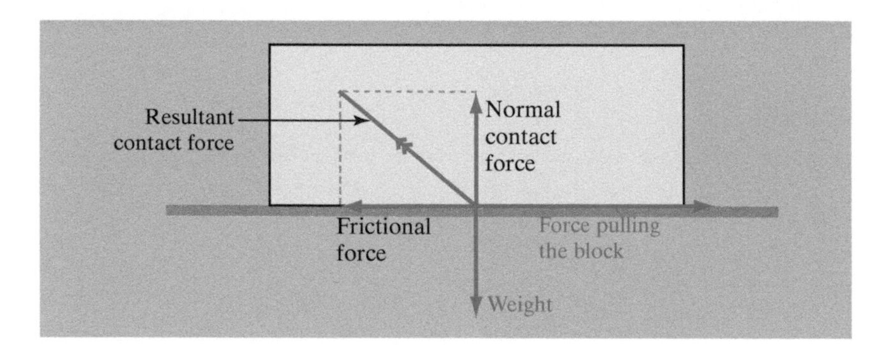

Fig 7
Forces on an object moving across a surface

Friction

A frictional force acts between two surfaces whenever there is relative motion between them, or when an external force is acting so as to slide them past each other.

Tension

An object is said to be in **tension** when a force is acting to stretch the object (Fig 8). Elastic materials, like ropes or metal cables, resist this stretching and exert a force on the bodies trying to stretch them.

Fig 8
The tension in the tow-rope, *T*, acts to pull the truck backwards and downwards, and to pull the car forwards and upwards.

Free body diagrams

The forces acting on a real object may be quite complex. A **free body diagram** is an attempt to model the situation so that we can analyse the effect of the forces. The free body diagram is used to show all the external forces that are acting on an object. Since forces are vector quantities they are represented by arrows, drawn to scale and acting in the correct direction.

Notes

Make sure that you only include forces acting on the object you are considering, e.g. the child on the slide. Don't confuse things by including forces that act on other objects, e.g. the slide, or by including internal forces, like the tension in the child's muscles.

Example

Draw a free body diagram for a child sliding down a playground slide.

Answer

Equilibrium

Once we have identified all the forces acting on an object and drawn a free body diagram, we can use vector addition to find the resultant force. If the resultant force is not zero then the object will accelerate in the direction of the resultant force. If an object is not accelerating, it is said to be in **equilibrium**.

Definition

An object is said to be in equilibrium if it is moving at constant velocity. This includes a velocity of zero, i.e. the object is stationary.

One of the conditions for equilibrium is that all the external forces that act upon the object must add up to zero. This means that if three forces are acting, the vector addition must form a closed triangle (Fig 9).

Fig 9
a + **b** + **c** must form a closed triangle if the body is to be in equilibrium.

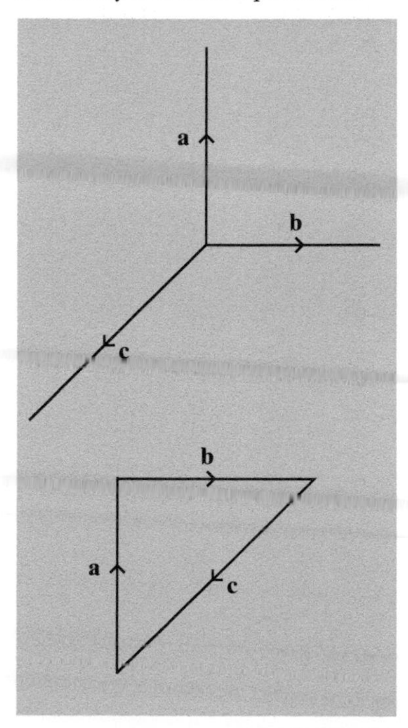

Example

A tightrope walker of mass 50 kg is standing on the middle of a tightrope. The rope makes an angle of 15° with the horizontal. Draw a free body diagram for the tightrope and find the tension in the rope. (To simplify the example, assume that the mass of the rope is negligible and take $g = 10\ \text{N kg}^{-1}$.)

Answer

Since the contact force between the rope and the person must balance the person's weight, the contact force, F, acting on the rope is 500 N. If the rope is to stay in equilibrium, the vector diagram must form a closed triangle.

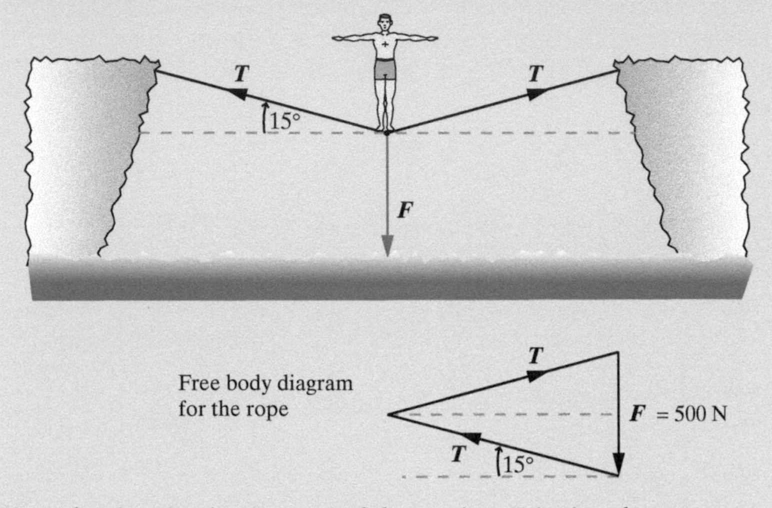

Since the situation is symmetrical the tension, T, is given by:

$$\frac{250}{T} = \sin 15°$$

$$T = \frac{250}{\sin 15°} = 966\ \text{N} = 970\ \text{N (to 2 s.f.)}$$

An alternative method of investigating these problems is to resolve all the forces into two perpendicular directions, for example horizontal and vertical. If the object is to be in equilibrium, two conditions must be satisfied:

- the sum of the horizontal components must be zero
- the sum of the vertical components must be zero.

Example

A car of mass 1200 kg is parked on a hill inclined at 20° to the horizontal. The maximum frictional force between the tyres and the road is 5000 N. Will the car remain in equilibrium?

Answer

First we need to resolve the car's weight W into components that are perpendicular and parallel to the slope. For equilibrium perpendicular to the slope, the normal contact force R is:

$$R = W \cos 20° = 12\,000 \times 0.94 = 11\,300\ \text{N}$$

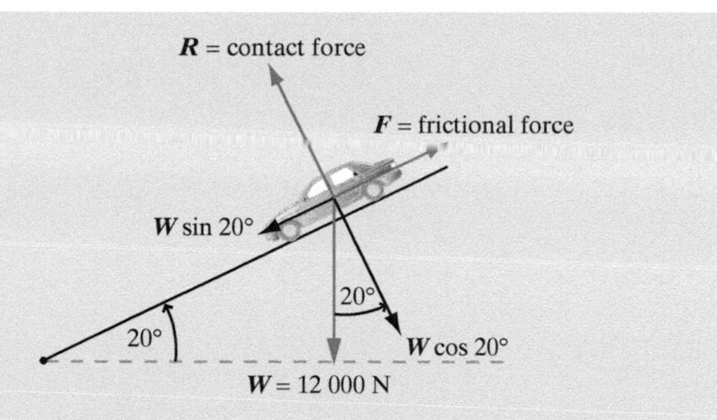

For equilibrium parallel to the slope the frictional force F is:

$F = W \sin 20° = 4100$ N

This is less than the maximum value; the car will remain in equilibrium.

3.4.1.2 Moments

Moment of a force

Forces can cause objects to accelerate in a straight line. They can also have the effect of turning or tipping an object. Even when an object is acted upon by two equal and opposite forces, it may not be in equilibrium. If the forces do not pass through a single point, the object will tend to rotate. For example, if you push a wardrobe at the top, and there is a large frictional force due to the carpet at the bottom, the wardrobe will tip rather than slide (Fig 10).

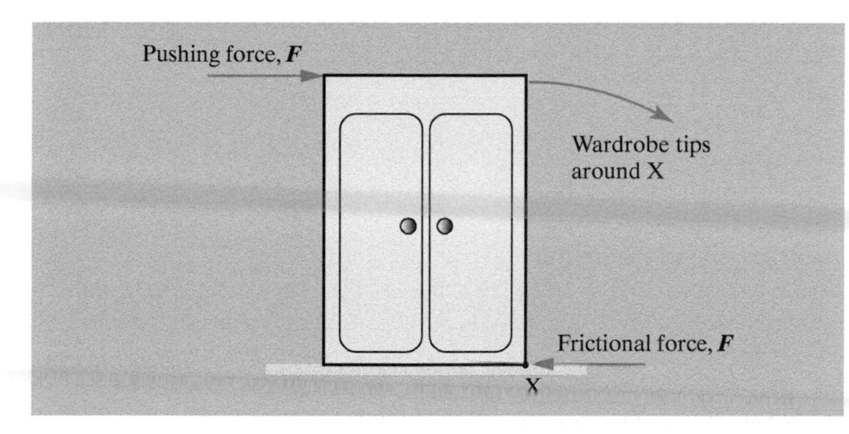

Fig 10
The moment about the point X causes the wardrobe to tip over.

Fig 11
The moment of force F about X is given by
moment = Fs

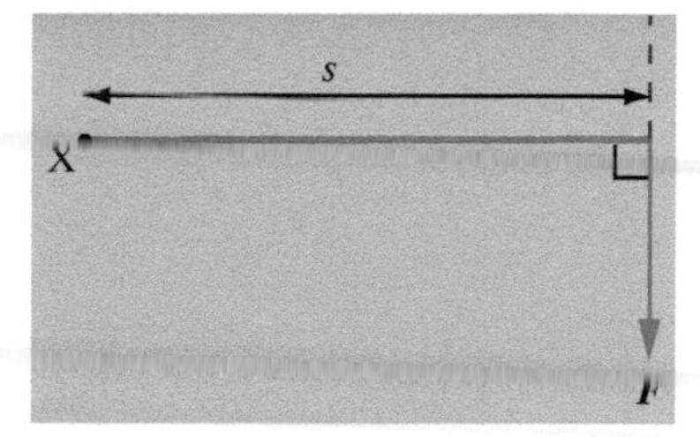

The turning effect of a force about a point is known as its **moment**. The moment of a force about a point depends on two things:

- the magnitude of the force
- the perpendicular distance from the line of the force to the point.

Definition

The moment of a force about a point is equal to the magnitude of the force, F, multiplied by the perpendicular distance of the force from the point, s.

moment (N m) = F (N) × s (m)

Notes

It is a common mistake to give the units for the moment of a force as newtons per metre (N/m). This is incorrect; the right units are N m.

The moment of a force is measured in newton metres.

You can increase the moment of the force you apply to a spanner on a nut by exerting a larger force, or by getting a longer spanner. When cycling a bicycle, the maximum turning effect is when the pedal crank is horizontal. When the pedal crank is vertical there is no moment, since the force passes through the pivot (Fig 12).

Fig 12
Moments on bicycle pedals

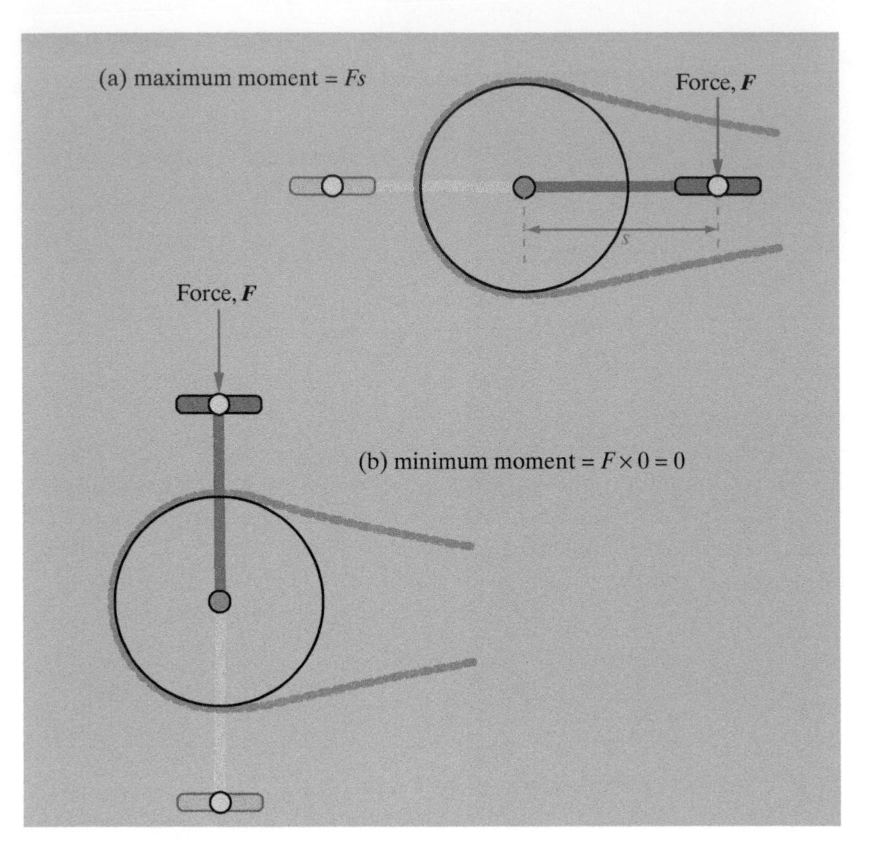

Essential Notes

It is important to note that it is the *perpendicular* distance from the pivot to the line of the force that is relevant. In Fig 13, where a vertical force is being used to lift a trap door, the moment is $Fs \cos \theta$.

Fig 13
Trap door hinged at X

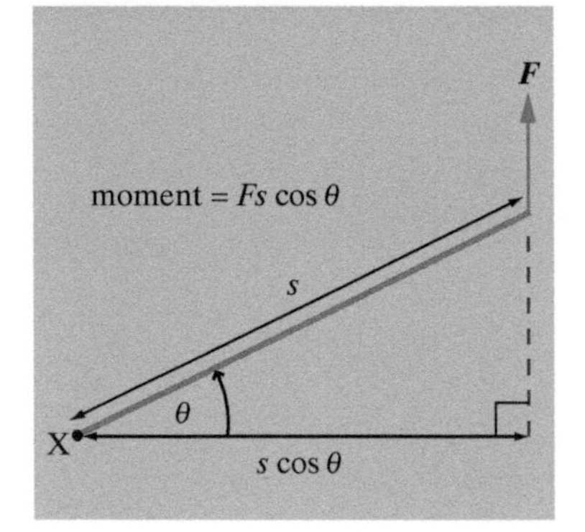

The principle of moments

There is a second condition that must be satisfied if an object is to be in equilibrium under the action of several forces. Not only must the vector sum of the forces be zero, but the sum of the moments about any point (taking their directions into account) must be zero. If this condition is not met, the object would rotate around that point.

Definition

*The **principle of moments** states that if an object is in equilibrium the sum of the moments about any point must be zero.*

Another way of putting this is to say that the sum of the moments which tend to turn the object anticlockwise must be equal to the sum of the clockwise moments. A familiar example of this is on a see-saw. A heavier child can be balanced by a lighter child if the lighter child sits further from the pivot (Fig 14).

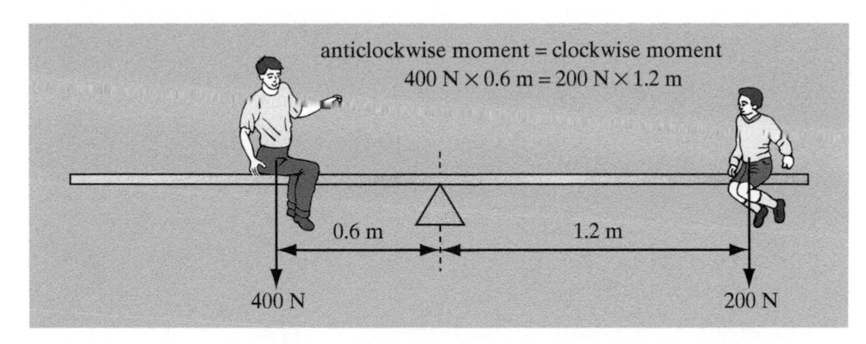

Fig 14
Equilibrium on a see-saw

The principle of moments can be applied to find the magnitude of unknown forces.

Example

A crowbar (lever) is used to lift a paving slab which weighs 300 N. The crowbar pivots at a point 0.20 m from the slab. How much force will it take to lift the slab, if the force is applied 1.2 m away from the pivot?

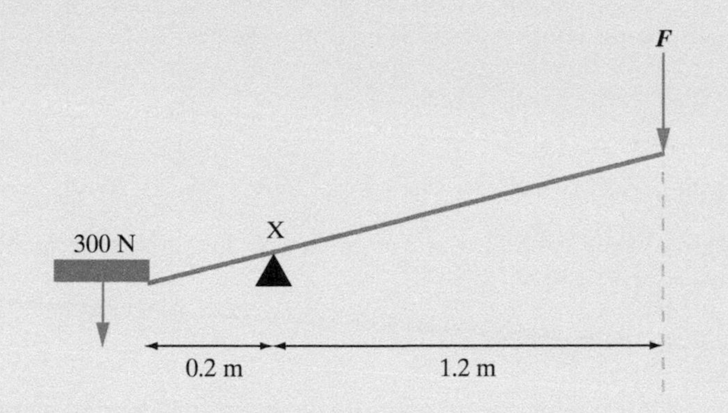

Answer

Taking moments about the pivot, X:

Anticlockwise moment = 300 N × 0.2 m = 60 N m

Clockwise moment = $F \times 1.2$ m = $1.2\,F$ N m

If the crowbar is in equilibrium these moments must balance:

$00 = 1.2\,F$, so $F = \dfrac{60}{1.2} = 50$ N

To lift the slab, the force must be just greater than 50 N.

Essential Notes

This lever gives you the ability to lift large weights with a smaller force. However, you will have to move your force much further than the slab will move. The **work** that you do will never be less than the **energy** gained by the slab.

Sometimes it is necessary to apply *both* the conditions for equilibrium in order to calculate all the forces in a situation.

- The vector sum of the forces must be zero.
- The sum of the moments about any point must be zero.

Example

Two people are carrying a 3 m long plank which has a mass of 20 kg. Andrew is holding the plank at one end, whilst Beryl is holding the plank 1 m from the opposite end. Calculate the forces that each person must exert. (Take $g = 10\ \text{N kg}^{-1}$ to simplify the example.)

> **Notes**
> **Always** draw a diagram showing the forces acting.

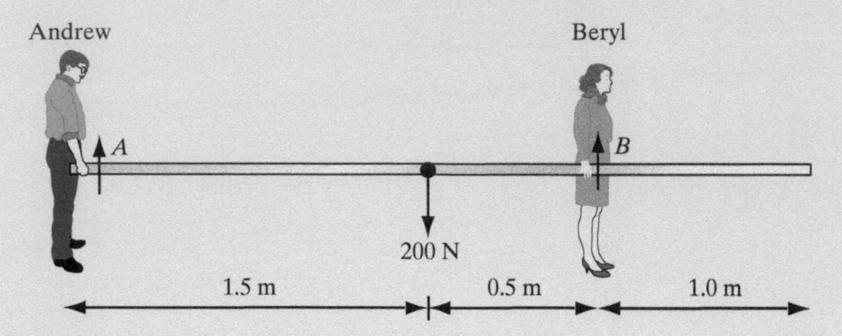

Answer

For equilibrium, the sum of the vertical forces are equal: $A + B = 200$ N

Taking moments about Andrew's end of the plank:

Clockwise $200 \times 1.5 = 300$ N m

Anticlockwise $B \times 2 = 2B$ N m

For equilibrium these must be equal: $2B = 300$ N m

So Beryl's force is 150 N. The rest of the 200 N plank is supported by Andrew, a force of 50 N.

> **Notes**
> Take care when choosing a point to take moments about. Choose a point that one of the unknown forces passes through so that this force has no moment about that point.

Couples

Two parallel forces which act in opposite directions will tend to make an object rotate. If these forces are equal in magnitude, they are known as a **couple**.

> **Definition**
> *The moment of a couple is Fs, where F is the magnitude of the forces and s is the perpendicular distance between the forces (see Fig 15).*

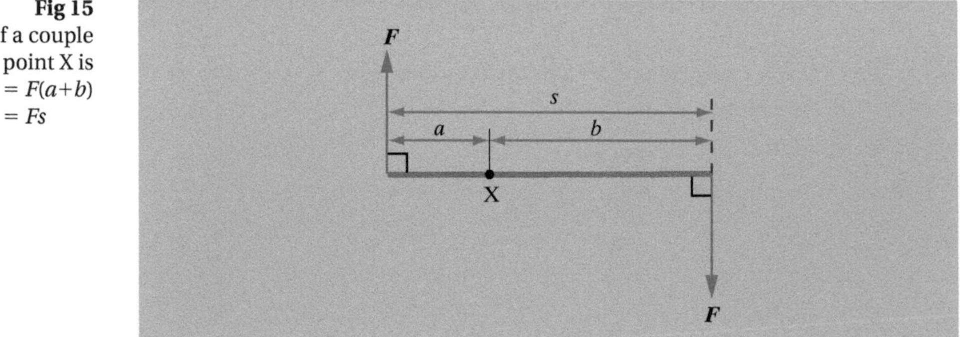

Fig 15
A couple. The moment of a couple about a point X is
$F \times a + F \times b = F(a+b) = Fs$

An example of a couple is in an electric motor where the force on each side of the coil is equal but opposite (Fig 16).

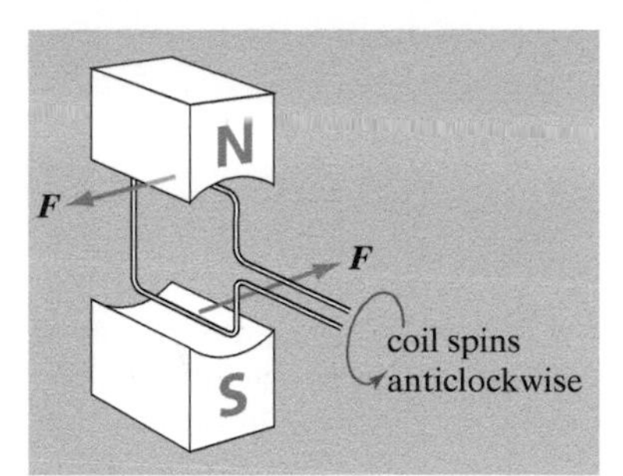

Fig 16
A couple in an electric motor

Centre of mass and centre of gravity

The total mass of a real object, for example a car, is the sum of the masses of all the individual parts that the object is made from. These parts will normally be different shapes and densities. However, we can find a point, called the **centre of mass**, which is the mean position of all the masses that make up the object.

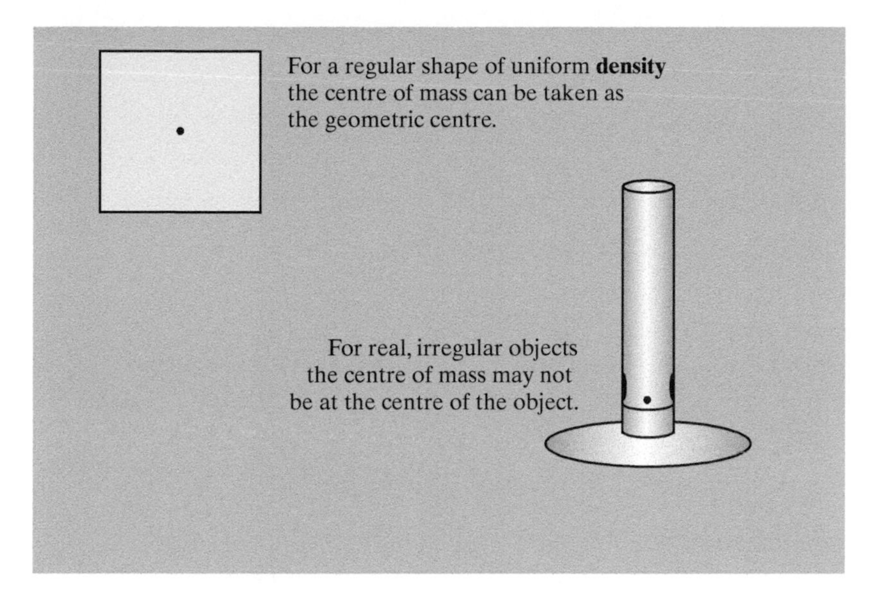

Fig 17

If we push the object along a line through the centre of mass, the object will accelerate in a straight line. But a force that does not pass through the centre of mass will cause the object to spin (Fig 18).

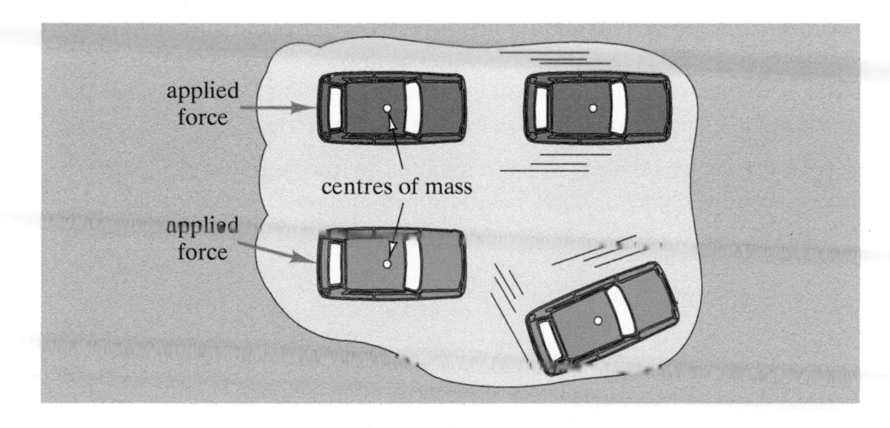

Fig 18
The applied force on the lower car does not pass through the centre of mass, so the car spins.

In a similar way, the weight of a real object is the sum of the weights of all its individual parts. We can treat the total weight as a single force acting at a point known as the **centre of gravity**. To make an object balance, you would need to support it above or below the centre of gravity.

Essential Notes

In practice, the centre of mass and the centre of gravity are the same point. They would only be at different positions if the strength of the gravitational field varied across the object.

3.4.1.3 Motion along a straight line

Displacement and distance

Displacement and distance are both measured in the same units, metres, but displacement, s, is a vector quantity that describes the *effect* of a journey rather than the total distance travelled. Distance travelled is a scalar quantity.

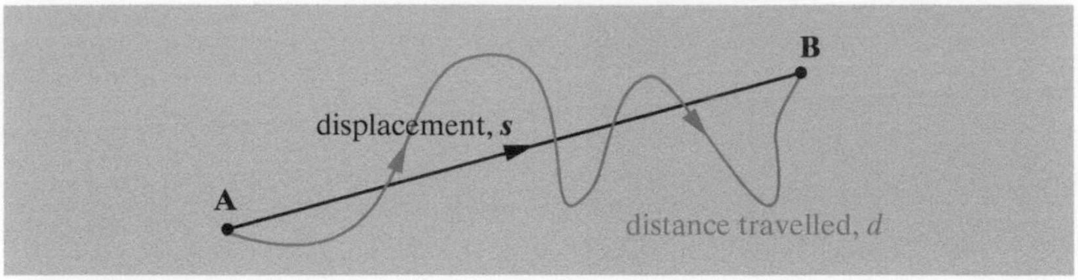

Fig 19
The vector s represents the displacement. This is the net effect of the journey, and in this case has a much smaller magnitude than the distance travelled.

Definition

Displacement, s, is the distance travelled in a given direction.

Speed and velocity

Speed is the distance covered in unit time. In SI units, this is metres in one second.

$$\text{speed (m s}^{-1}) = \frac{\text{distance travelled (m)}}{\text{time taken (s)}}$$

Speed is a scalar quantity which is measured in metres per second, or kilometres per hour. During a journey the speed may be changing. The average speed over the whole journey is given by:

$$\text{average speed} = \frac{\text{total distance covered}}{\text{total time taken}}$$

The speed at any given instant in the journey may be above or below the average speed. The speed at a certain time is known as the **instantaneous speed.** If we measure the distance covered in a very small time interval, Δt, the value for speed approaches the instantaneous value.

Essential Notes

The greek letter Δ (delta) represents a change in a physical quantity; Δs is a change in the displacement of an object.

Velocity is a vector quantity; it has a magnitude (measured in m s^{-1} or km h^{-1}) *and* a direction. Velocity is the speed in a given direction and is defined by:

$$\text{velocity (m s}^{-1}) = \frac{\text{displacement (m)}}{\text{time (s)}} \qquad v = \frac{\Delta s}{\Delta t}$$

Acceleration

Acceleration is the rate at which velocity changes.

$$\text{acceleration} = \frac{\text{change in velocity}}{\text{time taken for change}} = \frac{\Delta v}{\Delta t}$$

Essential Notes

A car driving round a roundabout may be travelling at a steady speed but it is constantly changing its velocity, because it is changing its direction.

Since the change in velocity is measured in m s^{-1}, and time is measured in seconds, acceleration is measured in m s^{-2}. Acceleration is a vector quantity and therefore takes place in a particular direction. Any change in velocity, either speeding up, slowing down or simply changing direction, is an acceleration.

Acceleration does not always take place in the same direction as the velocity. A ball thrown in the air which rises and then falls again is always accelerating downwards due to gravity (Fig 20).

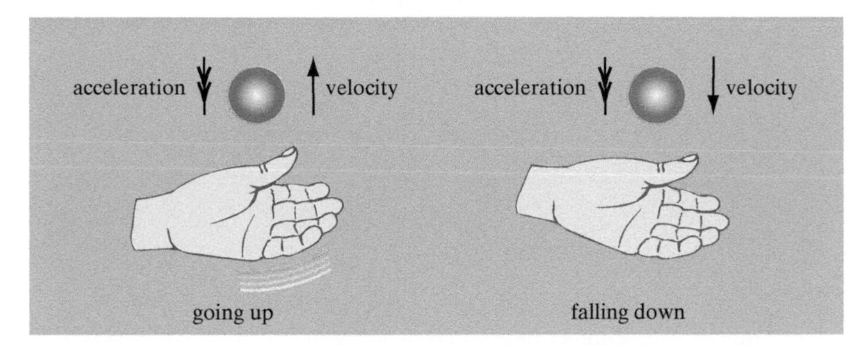

Fig 20
Acceleration and velocity of a ball thrown upwards (left) and falling back (right)

Example

A car goes from rest to 60 mph in 4.7 seconds. Calculate its acceleration. (1 mile = 1.6 km)

Answer

The car's final velocity is $\dfrac{(60 \times 1600)}{(60 \times 60)} = 27\,\text{m}\,\text{s}^{-1}$

acceleration $= \dfrac{\Delta v}{\Delta t} = \dfrac{27}{4.7} = 5.7\,\text{m}\,\text{s}^{-2}$

Displacement–time graphs

A journey can be represented by a graph showing displacement against time. The gradient of the graph represents the displacement in a certain time interval, which is the velocity. A straight line represents constant velocity.

Definition

The gradient of a displacement–time graph is the instantaneous velocity.

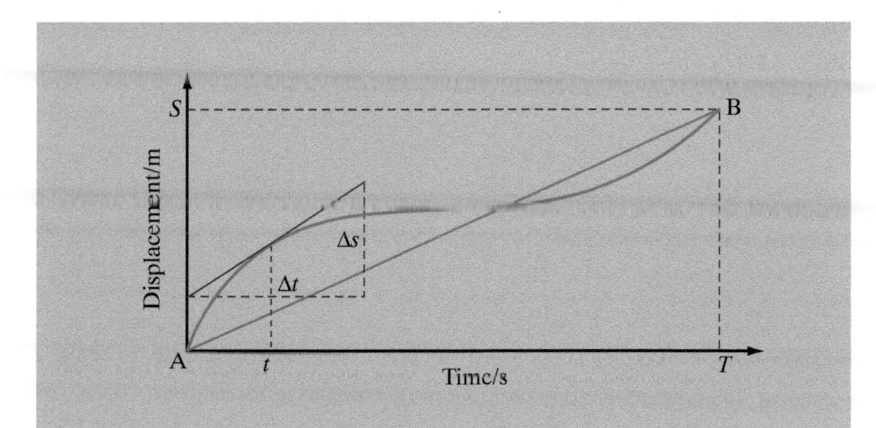

Fig 21
Instantaneous velocity and average velocity

(i) The instantaneous velocity at time t is the gradient of the curve at that point, $\Delta s/\Delta t$

(ii) The average velocity for the whole journey is the gradient of the straight line drawn from A to B, S/T

Example

The displacement–time graph below shows the motion of a car over 10 seconds. Describe what is happening at each stage of the journey.

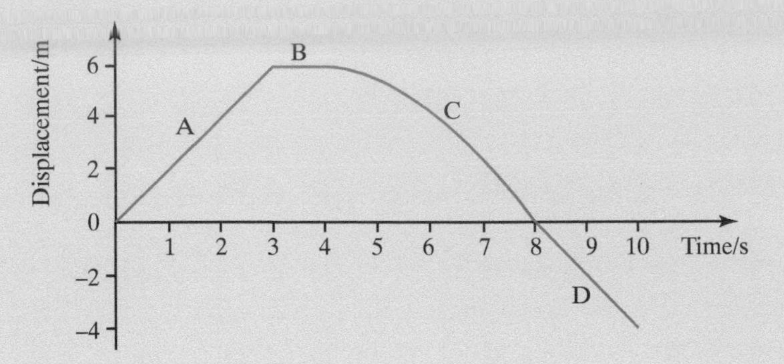

Answer

A Car moves with a constant velocity of $\frac{6}{3} = 2\,\text{m s}^{-1}$

B Between 3 and 4 seconds, the car is stationary.

C The car returns to its original position with non-uniform velocity. The car moves slowly at first and then more quickly.

D The car moves in the opposite direction (negative displacement) with a uniform velocity of

$-\frac{4}{2} = -2\,\text{m s}^{-1}$

The negative velocity means that the car is moving in the other direction.

Velocity–time graphs

A velocity–time graph for a journey can be used to calculate both the acceleration at any time and the total displacement. The acceleration can be found by calculating the gradient of the velocity–time graph. A straight line represents constant acceleration. If the gradient is negative, the object is slowing down, or speeding up in the opposite direction.

If the acceleration is changing, as in Fig 22, we can find its value at any instant in time by drawing a tangent to the graph at that time, and finding the gradient of that tangent.

The total displacement of a journey area is found by calculating the area below the velocity–time graph.

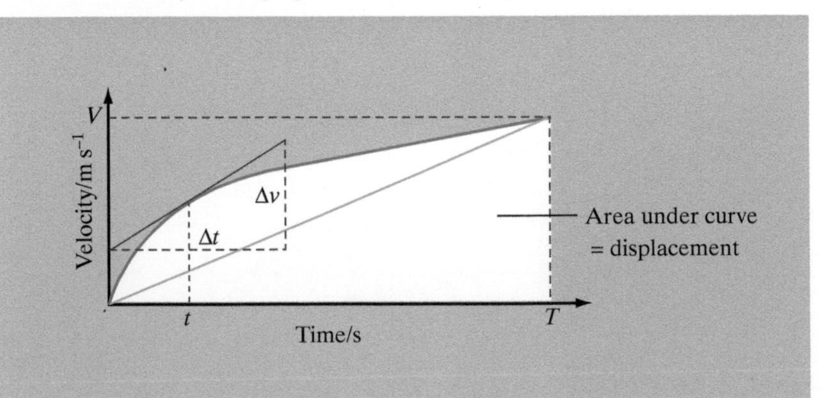

Fig 22
Velocity–time graph showing non-uniform acceleration. The instantaneous acceleration at time t is given by the gradient of the tangent to the curve, $a = \Delta v/\Delta t$.

The average acceleration over the total journey is the gradient of the straight line, V/T.

The displacement is the total area under the curve.

Example

The velocity–time graph below shows a person's journey on foot.
(i) Describe each section of the journey as fully as possible.
(ii) Calculate the displacement during the first 6 seconds.

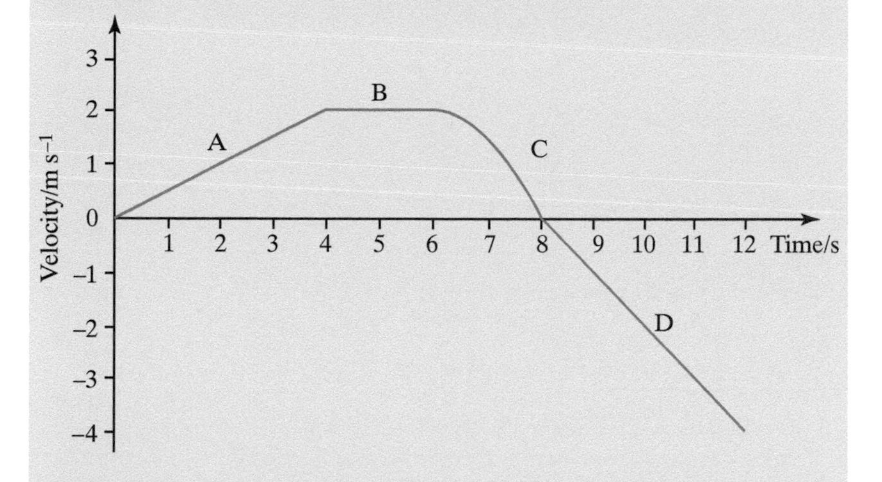

Answer

(i) A Uniform acceleration of $\frac{2}{4} = 0.5\,\text{m s}^{-2}$

B Constant velocity of $2\ \text{m s}^{-1}$

C Slowing down (negative acceleration) but at an increasing rate.

D Negative uniform acceleration: the person is now speeding up in the opposite direction.

Acceleration $= -\frac{4}{4} = -1\,\text{m s}^{-2}$

(ii) Displacement is the area under the graph $= \left(\frac{1}{2} \times 4 \times 2\right) + (2 \times 2)$

$= 8\ \text{m}$

Motion graphs for non-uniform acceleration

When a ball bounces vertically, the directions of the velocity and acceleration vectors continually change. In order to sketch the motion graphs for a bouncing ball, we first need to decide which direction is positive and make some simplifying assumptions:

- Define ground level as zero displacement.
- Define upwards as positive, down as negative.
- The ball falls, and bounces, vertically (in one dimension).
- The only forces acting on the ball are gravity and (briefly) the contact force of the ground. Air resistance is negligible.

Essential Notes

The directions we define as positive and negative are arbitrary. We can choose any reference level and sign convention, as long as we are consistent.

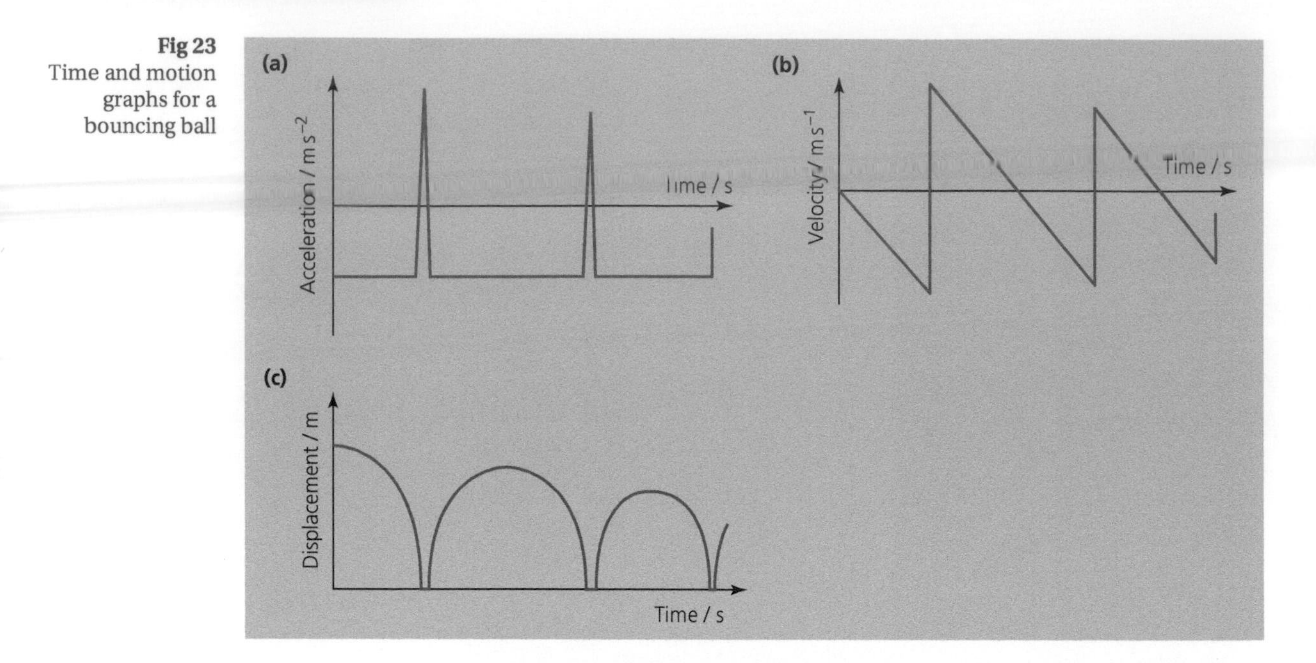

Fig 23
Time and motion graphs for a bouncing ball

Essential Notes

Even when the ball is travelling upwards, its acceleration is downwards.

1 **Acceleration–time graph (Fig 23 (a)).** Once the ball is released the only force acting on it is its weight. The ball has a constant acceleration of -9.81 m s^{-2} until it hits the ground. There is a brief resultant upward force, and acceleration, while the ball touches the ground. But as soon as the ball leaves the ground, the acceleration is -9.81 m s^{-2} once again. This is repeated for each bounce. The time between consecutive bounces gets shorter and the force exerted by the ground, and hence the upward acceleration, at each bounce also gets smaller.

2 **Velocity–time graph (Fig 23 (b)).** The velocity of the ball is initially zero and becomes more negative as the ball accelerates downwards. The gradient of the velocity–time graph is equal to the acceleration, and so is always -9.81 m s^{-2}, except briefly while the ball is in contact with the ground. The velocity is equal to zero at two points in each bounce: at the top of the bounce and during the contact with the ground.

3 **Displacement–time graph (Fig 23 (c)).** The velocity is the gradient of the displacement–time graph. The displacement starts at a positive value and drops to zero. The gradient starts at zero, as the ball is initially at rest, and then becomes increasingly negative as the ball accelerates downwards.

Equations of uniformly accelerated motion

We will just consider objects that move in a straight line with uniform acceleration. The five important variables that are used to describe this motion are shown in Table 2.

Table 2
Variables in uniform motion

Quantity	Symbol
displacement	s
initial velocity	u
final velocity	v
acceleration	a
time	t

There are a number of equations which link these variables together and describe uniformly accelerated, straight-line motion.

1 The definition of acceleration is:

$$\text{acceleration} = \frac{\text{change in velocity}}{\text{time}} \quad \text{or} \quad a = \frac{(v - u)}{t}$$

Rearranging this gives:

$$v = u + at$$

2 The definition of average velocity is:

$$\text{average velocity} = \frac{\text{displacement}}{\text{time}}$$

But if the velocity changes at a constant rate we can say that the average velocity is $(v + u)/2$.

So $$\frac{(v + u)}{2} \times t = s$$

or $$s = \tfrac{1}{2}(u + v)t$$

3 Equations 1 and 2 can be combined to give:

$$s = ut + \tfrac{1}{2}at^2$$

4 Equations 3 and 1 can be combined to eliminate t:

$$v^2 = u^2 + 2as$$

These four equations can be used to solve problems about motion.

Example

A sprinter accelerates from rest to 11 m s^{-1} in the first 4 seconds of a race. Assuming that his acceleration is constant, find the acceleration and the distance covered in the first 4 seconds.

Answer

$s = ?$ $u = 0\text{ m s}^{-1}$ $v = 11\text{ m s}^{-1}$ $a = ?$ $t = 4\text{ s}$

To find acceleration we can use the equation:

$$v = u + at;\ a = \frac{v - u}{t} = \frac{11}{4} = 2.75\,\text{m s}^{-2}$$

To find the displacement we can use $s = \tfrac{1}{2}(u + v)t$

$$s = \tfrac{1}{2} \times (0 + 11) \times 4 = 22\text{ m}$$

Acceleration due to gravity

A falling object accelerates towards the Earth because of the Earth's gravity. The **acceleration due to gravity** is independent of the object's mass. Experiments show that on Earth all objects, regardless of mass, accelerate under gravity at 9.81 m s^{-2}.

Example

Amy throws a ball vertically upwards with an initial velocity of 20.0 m s^{-1}. Calculate the maximum height reached.

Essential Notes

The acceleration due to gravity varies slightly from place to place on the Earth's surface; for example, it is greater at the poles than at the equator.

Answer

$v^2 = u^2 - 2as$

at the maximum height the vertical velocity is zero, so:

$0 = 400 - 2 \times 9.81 \times s$

$s = 20.4$ m

Example

Simon claims that the bubbles in his glass of fizzy drink are accelerating upwards at a constant rate (see fig). Simon estimates that there are 4 bubbles produced from the bottom of the glass every second.

(a) How could you find the acceleration?

(b) Use the diagram below to find the acceleration of the bubbles.

(c) How could Simon prove that the acceleration was constant?

Answer

(a) Measure the distance between two consecutive bubbles and divide by the time, 0.25 seconds in this case, to give the instantaneous velocity, v_1. Repeat this at a different point in the stream of bubbles to find v_2. Count the bubbles between v_1 and v_2, to calculate the time interval between each bubble. The acceleration is then $(v_2 - v_1)$ / time.

(b) $v_1 = 1$ cm / 0.25 = 4 cm s^{-1} and $v_2 = 2$ cm / 0.25 = 8 cm s^{-1}.
$a = (v_2 - v_1)$ / time = 8 – 4 / 1 = 4 cm s^{-2}.

(c) If the acceleration is constant then the equation $s = ut + ½ at^2$ applies. Assuming that $u = 0$, a graph of distance, s, against time-squared, t^2, should be a straight line.

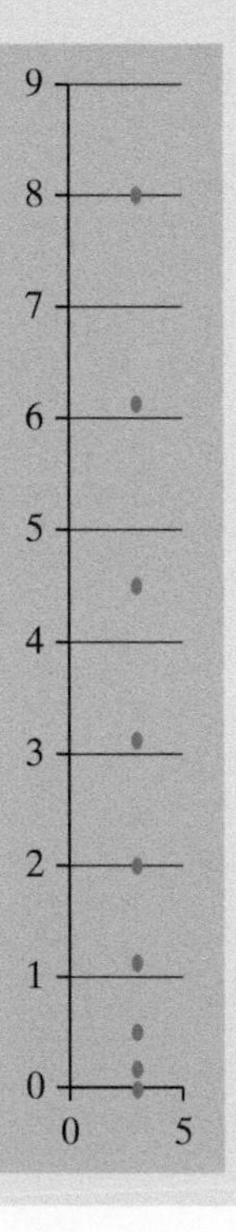

Example

A ball is thrown vertically downwards from a cliff 75.0 m high, with an initial velocity of 18.0 m s^{-1}. Calculate how much time elapses before the ball hits the ground. Take the acceleration due to gravity, $g = 9.81$ m s^{-2}

Using $s = ut + ½ at^2$

$75.0 = 18.0t + 4.91t^2$

$4.91t^2 + 18.0t - 75.0 = 0$

This is a **quadratic equation**, which we can solve using the quadratic formula.

So $t = -18.0 \pm \sqrt{(18.0^2 - 4 \times (-75.0) \times 4.91)} / 10.0$

which has two solutions, $t = -6.1$ s and $t = 2.5$ s

The positive answer is the one we need, $t = 2.5$ s.

Essential Notes

To solve a quadratic equation, rearrange it to the form $ax^2 + bx + c = 0$. The solution for x is given by $x = -b \pm \sqrt{(b^2 - 4ac)} / 2a$

3.4.1.4 Projectile motion

An object thrown through the air follows a parabolic path. Even though this is not a straight line, we can still use the equations of motion. This is because horizontal motion does not affect vertical motion (see Fig 24).

This means that a two-dimensional problem can be solved by treating it as two one-dimensional problems, i.e. keeping the horizontal and vertical motions separate.

Fig 24
An object dropped vertically and one thrown horizontally will fall at the same rate.

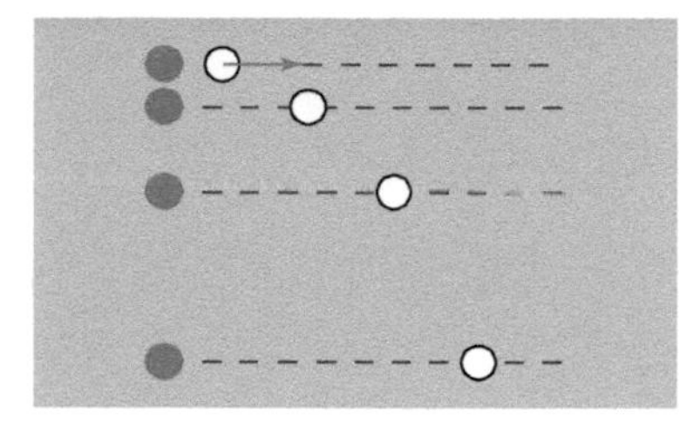

Example

An object is thrown with initial speed 20.0 m s^{-1} at an angle of 45° to the horizontal. Calculate its horizontal range and the maximum height it will reach.

Answer

First we resolve the initial velocity into horizontal and vertical components (Fig 25).

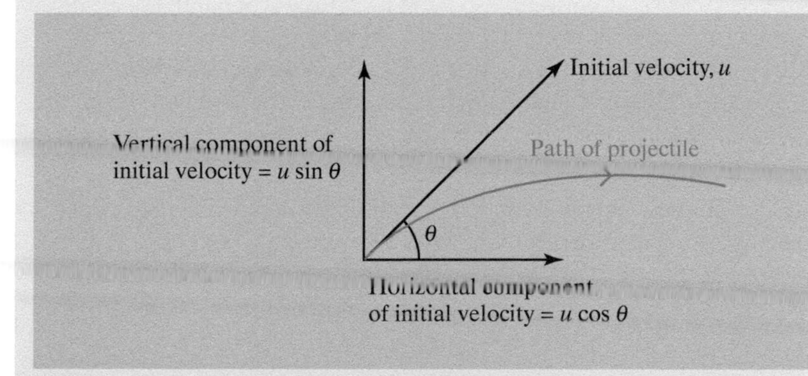

Fig 25 Components of velocity for a projectile

If air resistance can be ignored, there is no horizontal acceleration, so the initial and final horizontal velocities are the same.

Notes

Because displacement, velocity and acceleration are all vector quantities it is important to decide on a sign convention. When you start a problem you need to decide whether up is positive or negative and then stick to this throughout the problem.

The variables s, u, v, a and t for the horizontal and the vertical motion of the object are as shown in Table 3.

Horizontal motion	Vertical motion
s	s
$u = 20.0 \cos 45° \text{ m s}^{-1} = 14.1 \text{ m s}^{-1}$	$u = 20.0 \sin 45° = 14.1 \text{ m s}^{-1}$
$v = 20.0 \cos 45° \text{ m s}^{-1} = 14.1 \text{ m s}^{-1}$	$v = 0$ (halfway through the flight)
$a = 0 \text{ m s}^{-2}$	$a = -9.81 \text{ m s}^{-2}$ (down is negative)
t	t

Table 3 Variables in the horizontal and vertical motions of a projectile

Consider the vertical velocity first. When the object reaches its greatest height its vertical velocity will be zero: $v = 0 \text{ m s}^{-1}$.

We can find the greatest height s using $v^2 = u^2 + 2as$:

$$s = \frac{(v^2 - u^2)}{2a} = \frac{-200}{-19.6} = 10.2 \text{ m}$$

Using $v = u + at$, we can evaluate the time to reach the greatest height, t:

$$t = \frac{(v - u)}{a}$$

$$t = \frac{-14.1}{-9.81} = 1.44 \text{ s}$$

Then consider the horizontal motion. The time of flight is twice the time taken to reach the greatest height, so $t = 2 \times 1.44 = 2.88$ s. The horizontal range is given by:

$s = ut = 14.1 \times 2.88 = 40.6$ m

Essential Notes

Friction is not always a retarding force. It is often the force that propels things forward. Imagine trying to walk if there were no friction between your shoes and the ground, or turn a corner in a car that was driving on ice with bald tyres.

Friction

The frictional force, F, between two surfaces is proportional to the perpendicular (normal) contact force, N, between them (see 3.4.1.1). The frictional force also depends on the nature of the two surfaces; whether they are rough, like sandpaper, or smooth, like glass. The magnitude of the frictional force is given by: $F = \mu N$, where μ is known as the **coefficient of friction** between these surfaces.

This is the limiting, or static, friction. Dynamic friction, when the surfaces are sliding past each other, generally has a lower value. Friction acts to oppose relative motion between two surfaces. It acts to stop them sliding across each other.

Air resistance

Any object that is moving through a fluid is subject to a resistive force or **drag**. For example, any object moving through the atmosphere has to push the air out of the way, this gives rise to the drag force that acts to oppose relative motion between the object and the fluid (air).

Air resistance, or drag, has little effect on compact objects moving at low speeds, but the drag force increases with speed, and may eventually be equal to the force driving the object forward. When this happens the object will no longer accelerate.

For an object falling under gravity, the drag increases as the object accelerates, until the drag force is equal to the weight. When this happens, the object will fall at a constant velocity, known as the **terminal velocity** (Fig 26).

Essential Notes

The magnitude of the air resistance acting on an object depends on the area of the object and on the density of the air. The air resistance also increases as the relative speed between the object and the air increases. So, as you go faster, the force trying to stop you increases.

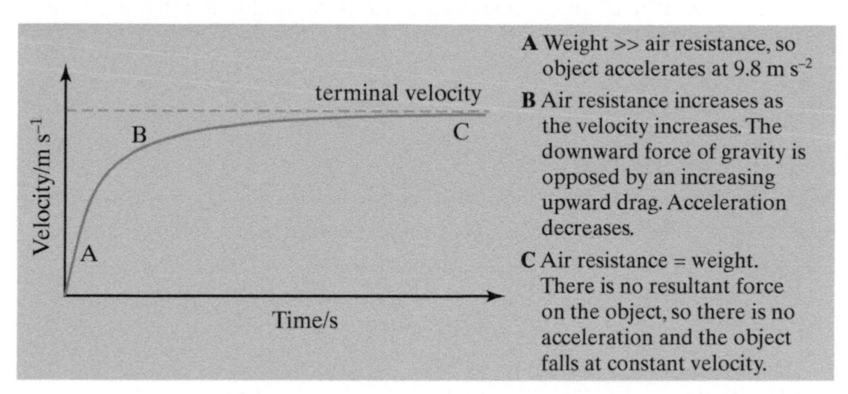

Fig 26
Velocity–time graph for an object falling in the Earth's atmosphere

Terminal velocity depends on the surface area of the object, as well as its mass, and on the density of the air. A falling feather floats down at less than 1 m s^{-1} whilst a free-fall parachutist may reach a terminal velocity of 65 m s^{-1} (close to 150 mph).

Essential Notes

Cars have a top speed because the drag due to air resistance and other resistive forces increase with speed. When the drag force is equal to the maximum driving force that a car can produce, the car will no longer accelerate. It has reached its top speed.

Effect of air resistance on projectile motion

The calculation of a projectile's trajectory on page 24 neglects the effect of air resistance. When air resistance is taken into account, both the horizontal range of a projectile and the maximum height that it reaches are reduced. The path is also no longer parabolic. This is because the horizontal component of the projectile's velocity does not remain constant.

The effect of air resistance on the trajectory of a projectile can be complicated. The shape and even the spin of a projectile will affect its path. For example, a cricket ball or football may swerve, or a golf ball may hang in the air longer than expected. If the object is shaped in a suitable way, for example an aircraft wing, then the overall effect can be an upward force, known as **lift**.

3.4.1.5 Newton's laws of motion

Newton's laws of motion were published in his *Principia* in 1687. His work on mechanics built on that of Galileo.

Newton's First Law

Definition

Newton's First Law of motion states that every object will continue to move with uniform velocity unless it is acted upon by a resultant external force.

Fig 27
If the lift is to keep moving down at a steady velocity, by Newton's First Law its weight must be balanced by the friction, F, and the tension in the lift cable, T. For equilibrium, $W = F + T$

Essential Notes

Momentum can also be given in units of newton seconds (N s); this is exactly equivalent to kg m s^{-1}.

This law expresses the idea that objects will stay at rest, or keep moving in a straight line at a steady speed, unless an external force acts on them. The law restates Galileo's law of inertia. **Inertia** is the reluctance of an object at rest to start moving, and its tendency to keep moving once it has started. This law isn't immediately apparent on Earth. If you push an object, it doesn't keep going in a straight line for ever, because on Earth it is difficult to avoid external forces like gravity or friction. In space, well away from any gravitational attractions, objects just keep moving in a straight line, at a steady speed.

Another way of stating Newton's First Law is to say that an object will remain in equilibrium, unless it is acted upon by an external force.

Newton's Second Law

Newton's Second Law relates to an object's **momentum**. The momentum of a moving object is defined as its mass multiplied by its velocity (see pages 29–34 for more about momentum). Momentum is a vector quantity, its direction the same as its velocity. It is given the symbol p. The units of momentum are those of mass × velocity, kg m s^{-1}.

Definition

Momentum (kg m s^{-1}) = *mass* (kg) × *velocity* (m s^{-1}) *or* $p = mv$

The momentum of a moving object can be thought of as a measure of how difficult it is to stop the object. For example, a heavy lorry, mass 40 tonnes, could be travelling along a motorway at 25 m s^{-1}. Its momentum would be:

$p = 40 \times 10^3 \text{ kg} \times 25 \text{ m s}^{-1} = 1.0 \times 10^6 \text{ kg m s}^{-1}$

Compare this to the momentum of a person running at top speed:

$p = 80 \text{ kg} \times 10 \text{ m s}^{-1} = 800 \text{ kg m s}^{-1}$

Using the concept of momentum, Newton's Second Law states the effect of a force on the motion of an object.

Definition

Newton's Second Law of motion states that the rate of change of an object's linear momentum is directly proportional to the resultant external force. The change in momentum takes place in the direction of the force.

Force, F, is proportional to the change in momentum, p, divided by the time taken for the change, Δt.

$$F \propto \frac{\Delta p}{\Delta t} \quad \text{or} \quad F \propto \frac{\Delta(mv)}{\Delta t}$$

This can be written $F = \dfrac{k\,\Delta(mv)}{t}$, where k is a constant.

If the mass of an object does not change then m is constant and:

$$F = \frac{km\,\Delta v}{\Delta t}$$

but $\frac{\Delta v}{\Delta t}$ = acceleration, a, so

$$F = kma$$

The unit of force, the newton (N), is defined to be equal to 1 when $m = 1$ kg and $a = 1$ m s^{-2}.

Definition

*The SI unit of force is the **newton**, N. One newton is the force that will accelerate a mass of 1 kg by 1 m s^{-2}.*

This means that in $F = kma$, $k = 1$ and we can write:

$$F = ma$$

force (N) = mass (kg) × acceleration (m s^{-2})

Essential Notes

$F = ma$ holds only for a constant mass.

Example

The total mass of a lift and its passengers is 1000 kg. The tension in the cable pulling the lift up is 12 000 N. Find the acceleration of the lift. (Ignore friction and take $g = 10$ N kg^{-1} to simplify the example.)

Answer

Upward force = 12 000 N

Downward force = mg = 1000 kg × 10 N kg^{-1} = 10 000 N

Resultant upwards force = 2000 N

Using $F = ma$, $a = \frac{2000}{1000} = 2.0\,\text{m s}^{-2}$

Notes

Remember to find the resultant force first, then use Newton's Second Law to find the acceleration.

Newton's Second Law can be used to explain the action of car safety features such as seat belts, air bags and crumple zones. The force acting on a passenger in a collision is equal to their rate of change of momentum:

$$F = \Delta(mv) / \Delta t$$

The change of momentum is fixed by the initial speed of the car and the mass of the passenger. The only way to reduce the force acting on the passenger is to increase the time taken for them to come to a stop. All the safety features mentioned above are designed to increase the stopping time.

In some cases, such as rockets and jets, mass cannot be treated as constant. It is then necessary to think of force as the rate of change of momentum, and to apply Newton's Second Law in the form:

$$F = \frac{\Delta(mv)}{\Delta t}$$

Notes

You will not be asked a question involving changing mass at AS. This will be dealt with further in AQA Physics A-level Year 2.

Newton's Third Law

Definition

Newton's Third Law of motion states that if an object, A, exerts a force on a second object, B, then B exerts an equal but opposite force back on object A.

Essential Notes

This law is sometimes written as 'every action has an equal but opposite reaction', but you need to be careful with this statement. Remember that *the forces act on different bodies.*

This law means that force between two bodies always acts equally on both objects, though in opposite directions. When we say that someone weighs 500 N, we mean that the gravitational attraction of the Earth on the person is 500 N. The person also attracts the Earth upwards with a force of 500 N. When a car exerts a force on the ground through the friction between its tyres and the road surface, the car pushes the ground backwards, whilst the ground pushes the car forwards. The two forces involved in Newton's Third Law never cancel each other out, because they act on different bodies.

The two forces in Newton's Third Law are always of the same type, e.g. gravitational.

Example

A man stands still on the surface of the Earth. Draw the forces acting on the man and on the Earth and explain which forces are equal to each other and why.

Answer

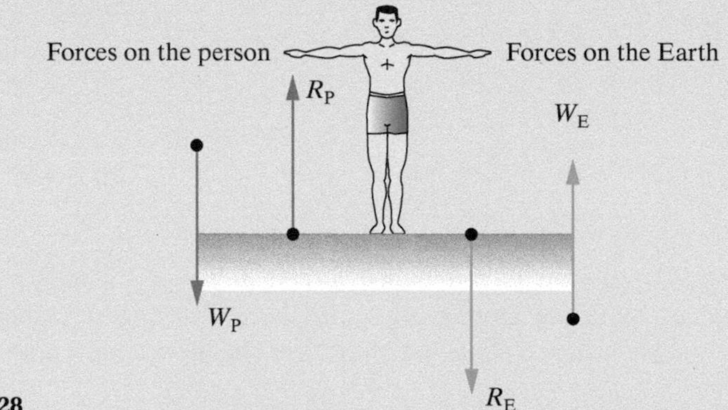

Fig 28

$W_p = W_E$	By Newton's Third Law, the weight of the person is equal to the gravitational attraction of the person acting on the Earth.
$R_p = R_E$	By Newton's Third Law, the contact force of the Earth pushing on the person is equal to the contact force of the person pushing on the Earth.
$W_p = R_p$	By Newton's First Law, because the person is in equilibrium, the person's weight is balanced by the contact force of the Earth.
$W_E = R_E$	By Newton's First Law, because the Earth is in equilibrium, the gravitational attraction of the person is balanced by the contact force of the person on the Earth.

3.4.1.6 Momentum

Conservation of linear momentum

When there are no external forces acting on an object its linear momentum does not change. This same principle can be extended to a system of several objects. The **conservation of linear momentum** is an important principle in physics which can be stated as follows.

Definition

The total linear momentum of a system is constant provided that there is no external resultant force acting.

It isn't immediately obvious that this is true. A car standing at traffic lights has no linear momentum, yet a few seconds later it has gained momentum as it pulls away from the lights. This does not contravene the conservation of momentum, since if we are just thinking about the car, the friction of the road on the car is an external force and conservation of momentum does not apply. If we consider the car and the Earth as the 'system' there are no external forces acting, so conservation of momentum must apply. As the car gains momentum in a forward direction, the Earth gains momentum in the opposite direction.

Similarly, when you jump into the air, you are moving the Earth a little. The Earth has to acquire a momentum that is equal but opposite to your upwards momentum. Since your mass is much less than the Earth's, the Earth's velocity will be much smaller than yours.

Collisions

The conservation of momentum is often applied to situations where two objects collide. If we can ignore external forces, such as friction, the total momentum before the collision must be equal to the total momentum afterwards.

Suppose car A has a mass of 1500 kg and is moving at 20 m s^{-1} when it collides into car B, which has a mass of 1000 kg and is moving in the opposite direction at 10 m s^{-1}. After the collision the two cars stick together. We can use the conservation of momentum to calculate the velocity after the collision.

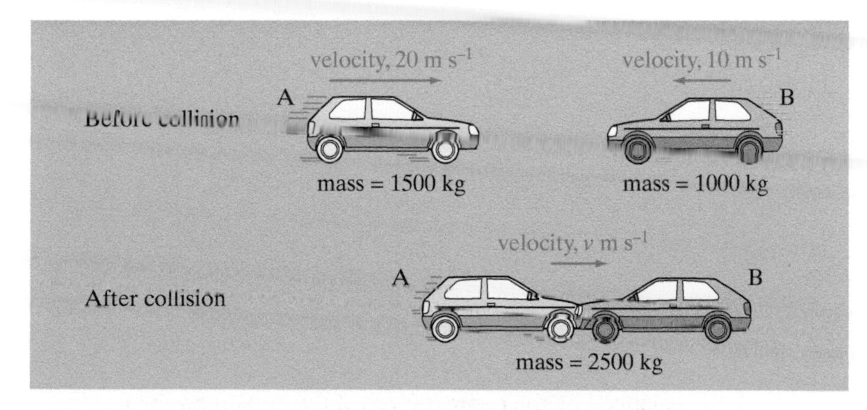

Fig 29
Conservation of momentum in a car collision

Notes

Remember that momentum is a vector quantity. If the momentum in one direction is positive, the momentum in the opposite direction must be negative.

Momentum of car A = 1500 kg × 20 m s^{-1} = 30 000 kg m s^{-1}

Momentum of car B = 1000 kg × (−10 m s^{-1}) = −10 000 kg m s^{-1}

Before the collision the total momentum = 20 000 kg m s^{-1}

After the collision the mass of the vehicles is 2500 kg and their velocity is v m s^{-1}.

The momentum after the collision = 2500 × v

If we can ignore any external forces the momentum before and after the collision must be the same:

$$20000 = 2500 \times v$$

$$v = \frac{20000}{2500} = 8\,\mathrm{m\,s^{-1}}$$

Essential Notes

Linear momentum is conserved in both elastic and inelastic collisions, provided there is no resultant external force.

Elastic and inelastic collisions

Collisions are classified as **elastic** or **inelastic**.

Definition

In an elastic collision there is no loss of kinetic energy.

If a collision is elastic, the total kinetic energy is the same before and after the collision. In an inelastic collision, kinetic energy is transferred to energy in different forms, such as heat. All collisions between everyday objects are inelastic – some energy is always transferred to other forms. Elastic collisions can take place between the molecules in a gas or between subatomic particles.

Elastic and inelastic collisions can be investigated in a school laboratory using gliders which run on an air track. The track has air blown through it so that the gliders rest on a cushion of air, eliminating friction between the glider and the track. If the air track is carefully levelled so that the glider moves horizontally, gravity will have no effect on the motion and we can say that there are no significant external forces acting on the glider. Light gates are used to time the glider's motion and a datalogger to calculate the velocity of the gliders so that the motion can be analysed (Fig 30).

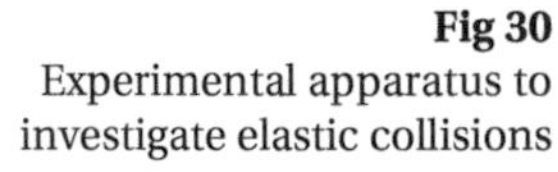

Fig 30
Experimental apparatus to investigate elastic collisions

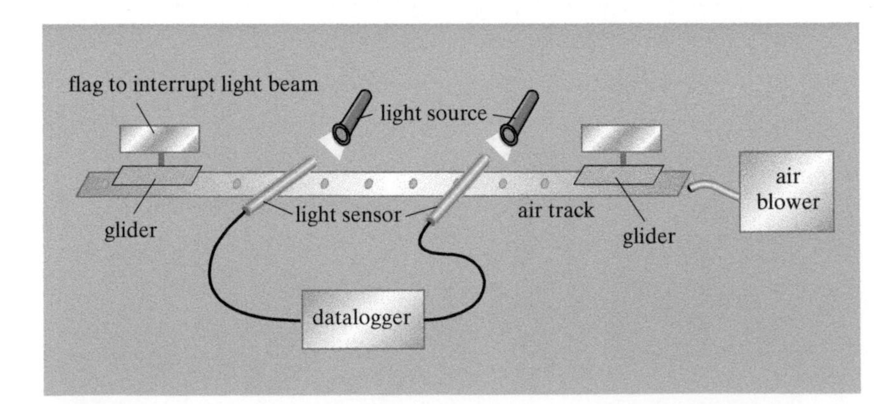

Explosions

The operation of a rocket engine can be explained by the conservation of momentum. The rocket engine propels hot gases out of the rear. The rocket must gain an equal momentum in the opposite direction (Fig 31).

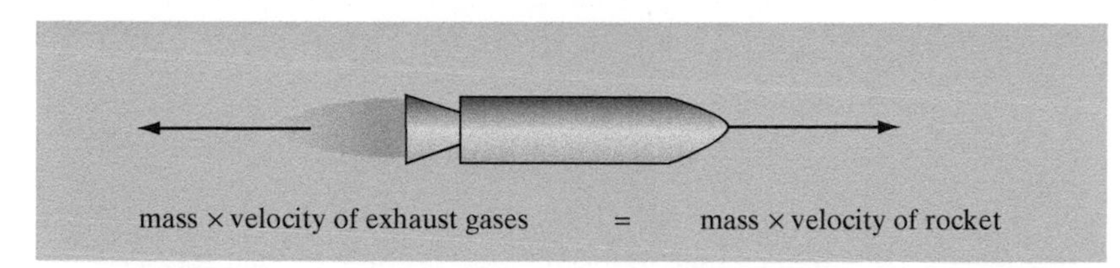

Fig 31
Conservation of momentum in a rocket

Essential Note

We need to be careful when considering problems involving rockets. As the rocket burns fuel, its total mass reduces.

Example

A spacecraft of mass 20 000 kg which is at rest fires its rockets. The exhaust gases are expelled at a rate of 100 kg s^{-1} and at a speed of 1000 m s^{-1}. If the rocket is fired for 10 seconds, what will the final velocity of the spacecraft be?

Answer

Total momentum before firing is zero. Therefore total momentum afterwards is also zero; the 'reverse' momentum of the exhaust gases must be balanced by the forward momentum of the rocket.

In 10 s the mass of exhaust gases is 100 kg s^{-1} × 10 s = 1000 kg

Total momentum of gases = mv = 1000 kg × 1000 m s^{-1}
= 1 000 000 kg m s^{-1}

Momentum of spacecraft must also be 1 000 000 = 19 000 × v, so v = 53 m s^{-1} (to 2 s.f.)

Note that the rocket's mass has decreased by 1000 kg

Notes

It is a common misconception that rockets work by pushing against the Earth. This can't be true, because rockets work out in space where there is no external object to push against. Rocket engines push against the exhaust gases, which push back with an equal but opposite force on the rocket.

This is an example of Newton's Third Law.

The conservation of momentum also explains why a gun recoils when it fires a bullet. Before the gun is fired, the initial momentum of the system, that is the gun and the bullet, is zero. The final momentum of the system must also be zero. This means that any forward momentum acquired by the bullet must be balanced by the momentum of the gun recoiling in the opposite direction.

The same considerations apply to all types of explosions, such as a bomb exploding or the emission of an alpha or beta particle from a **nucleus** (Fig 32).

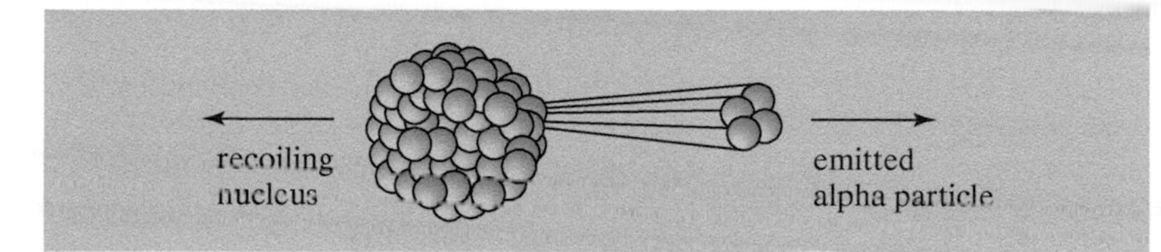

Fig 32
The momentum of the emitted alpha particle is equal but opposite to the momentum of the recoiling nucleus. Since the nucleus has a greater mass, it will recoil at a lower velocity than that of the alpha particle.

Force, momentum and impulse

Remember from Newton's Second Law of motion that an equation for force is

$$F = \frac{\Delta(mv)}{\Delta t}$$

If we rearrange this equation we obtain

$$F\,\Delta t = \Delta(mv)$$

and we see that the total change in momentum is equal to the product $F\,\Delta t$. This product, the force multiplied by the time over which it acts, is known as the **impulse**. It has units of newton seconds, N s.

Definition

***Impulse** is the magnitude of a force multiplied by the time for which it acts: the impulse of a force F that acts for a time Δt is F Δt.*

The impulse is useful for finding the change in momentum of an object.

Example

A golf club strikes a golf ball, of mass 45 g, with an average force of 5 kN. The contact between the club and the ball lasts for 0.5 ms. Find the velocity at which the golf ball leaves the club.

Answer

The impulse of the force is $F\,\Delta t = 5000 \times 0.5 \times 10^{-3} = 2.5$ N s

This is equal to the change in momentum of the golf ball.

$\Delta(mv) = 2.5$ N s

As the mass is constant,

$m\,\Delta v = 2.5$ N s

so

$$\Delta v = \frac{2.5}{0.045} = 55.6\ \text{m s}^{-1}$$

This is the velocity at which the ball leaves the club, as initially it is at rest.

In many cases the force is not constant, but varies with time. The total impulse, and hence the change in momentum, is then given by the area under the force–time graph. For example, when a tennis racket hits a tennis ball, both objects are squashed. The ball and the racket act like springs. The force is initially small, increases to a peak value and then decreases again as the racket and ball return to their original shapes (Fig 33).

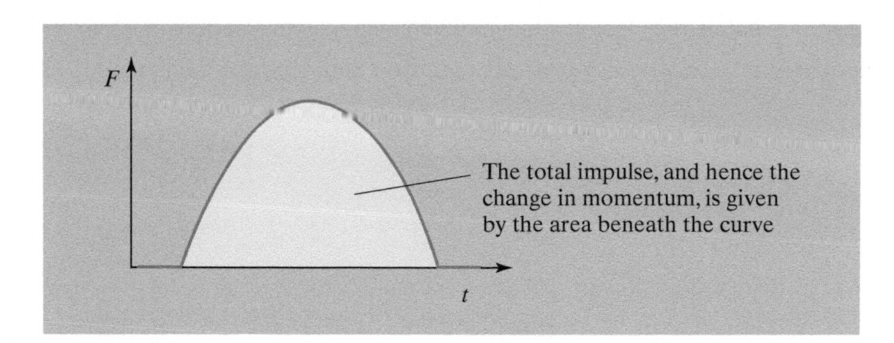

Fig 33
The force between a tennis racket and ball during a shot, against time

Water jets

In some cases, such as rockets and jets, mass cannot be treated as constant. It is then necessary to think of force as the rate of change of momentum, and to apply Newton's Second Law in the form:

$$F = \frac{\Delta(mv)}{\Delta t}$$

Example

A pressure washer can eject 1.5×10^{-2} m^3 of water per minute at a speed of 52 m s^{-1}. The water jet hits a wall and comes to rest. Find the force exerted by the jet on the wall. (Density of water = 1000 kg m^{-3})

Answer

In 1 second, the pressure washer ejects $\frac{1.5 \times 10^{-2}}{60} = 2.5 \times 10^{-4}$ m^3 litres of water.

This is a mass of $2.5 \times 10^{-4} \times 1000 = 0.25$ kg s^{-1}.

The water is moving at 52 m s^{-1} so the momentum lost by the water in 1 second is:

$p = mv = 0.25 \times 52 = 13\,\text{kgms}^{-1}$

Since force in newtons is equal to the change of momentum in 1 second, the force is 13 N.

Inelastic collisions

The law of conservation of momentum is obeyed by all collisions. Some collisions, referred to as **elastic**, also conserve kinetic energy. For example, the collisions between gas molecules.

Collisions between macroscopic objects are always **inelastic** to some extent; momentum is conserved but kinetic energy is not. Some of the kinetic energy is transferred by heating the surroundings. This is why a bouncing ball rises less after each bounce.

It is important that some collisions are inelastic. Crumple zones and air bags in cars are designed to dissipate kinetic energy, and to reduce the maximum force on passengers. The instantaneous force acting on a passenger during the collision is equal to the rate of change of momentum at that time. If we consider the whole collision:

average force = total change in momentum / time

The change in momentum is fixed by the speed and the mass of the vehicle. The only way to reduce the force on the passengers is to increase the time of the collision.

Example

In an adventure movie, a stuntman has to fall from a second-floor window. Stacks of cardboard boxes are used to break his fall.

(a) The collision between the stuntman and the boxes is inelastic. Explain what this means and why it is important in this case.

(b) Explain, with reference to Newton's second law, how the cardboard boxes help.

(c) During the scene described above, the stuntman's collision with the cardboard boxes lasts for 1.0 seconds. Estimate the average force on the stuntman.

Answer

(a) Inelastic means that kinetic energy is not conserved. If the collision were elastic, the stuntman would bounce off without slowing down.

(b) Newton's Second Law says that the force = total change in momentum / time. Therefore increasing the duration of the collision will reduce the force on the stuntman.

(c) Estimate height of second floor window = 10 m.

Speed at which stuntman reaches the ground is given by $v^2 = u^2 + 2as$, since $u = 0\ \mathrm{m\ s^{-1}}$ and $a = 9.8\ \mathrm{m\ s^{-2}}$, $v = \sqrt{2 \times 9.8 \times 10} = 14\ \mathrm{m\ s^{-1}}$

Initial momentum = mass of stunt man × velocity = 80 kg × 14 m s^{-1} = 1120 kg m s^{-1}.

Final momentum = 0, so the average force 1120 / 1 = 1120 N

Fig 34

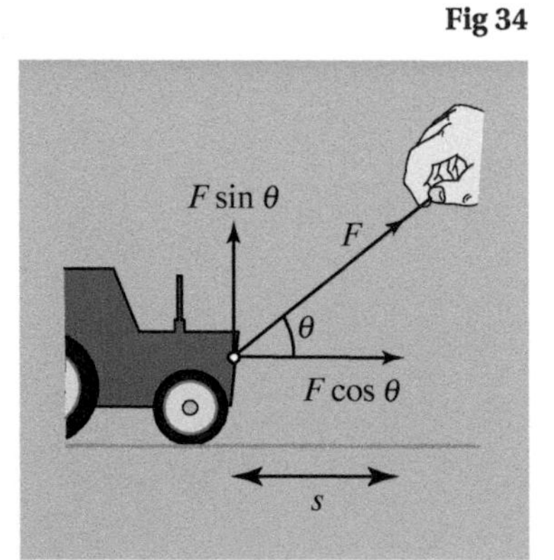

3.4.1.7 Work, energy and power

Work

Work is done whenever a force moves through a distance in the direction of the force. A force that does not move does no work. Sometimes the motion is not in the same direction as the applied force, e.g. when a toy tractor is being pulled along by a string (Fig 34).

The force, *F*, can be resolved into two components: parallel to the motion and perpendicular to it. There is no movement in the direction of the perpendicular component, $F \sin \theta$, and so it does no work. All the work done is by the component that is parallel to the displacement, $F \cos \theta$.

Definition

The work done is equal to the force multiplied by the distance through which the force moves, in the direction of the force.

work done, W (J) = F (N) × s (m) × cos θ

Work is measured in joules (J). One joule of work is done whenever a force of 1 newton moves through 1 metre.

Example

A barge is pulled along a canal by a tow-rope (Fig 35). The tension in the tow-rope is 1000 N and it makes an angle of 20° with the canal. Find the work done in towing the barge 100 m forwards along the canal.

Answer

work done = 1000 × 100 × cos 20° = 94 kJ (to 2 s.f)

Fig 35

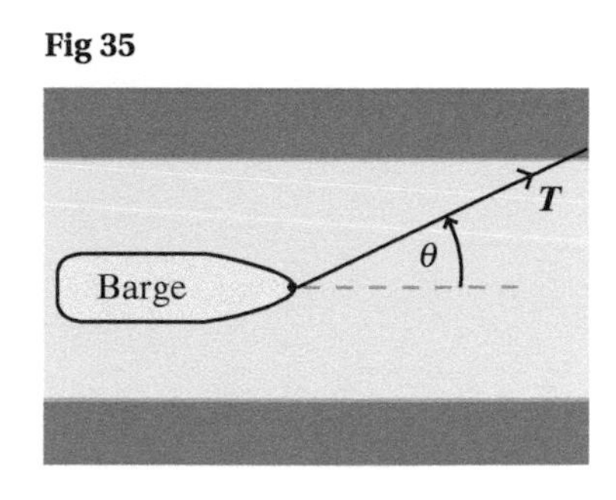

Energy and power

Definition

***Energy** is the ability to do work.*

Energy can appear in many different forms. It may be classified as electrical, chemical, kinetic (movement), gravitational potential, electric potential, thermal, electromagnetic (e.g. light), nuclear or sound. Each of these forms of energy can be transferred so that the final effect is doing some work, such as the lifting of a weight. For example, electrical energy can be transferred as kinetic energy by an electric motor. This kinetic energy can be transferred as **gravitational potential energy** as the axle of the motor winds up a string and lifts a weight. At every stage some energy is transferred as random thermal energy, or heat. In this way, energy tends to become less concentrated and less useful. All forms of energy are measured in the same unit, the joule.

The rate at which a device can transfer energy is known as its **power**. Power is measured as the number of joules per second that are transferred. The SI unit of power is the **watt**.

Definition

***Power** is the rate at which energy is transferred. A power of 1 watt means that 1 joule of energy is transferred every second.*

A powerful electric motor can transfer thousands of joules of electrical energy into kinetic energy per second and its power would be given in kilowatts (kW).

Example

An electric shower has a power rating of 7 kW. If the shower runs for 10 minutes, how many joules of energy will have been transferred?

Answer

The shower transfers electrical energy into thermal energy in the water at a rate of 7000 joules per second. In 10 minutes, this is $10 \times 60 \times 7000 = 4\,200\,000$ J or 4.2 MJ

Since energy can result in work being done, power can also be thought of as the rate at which work is done.

Definition

Power, P, is the rate at which work is done, $P = \frac{\Delta W}{\Delta t}$

Essential Notes

This is the output power of the person. The person would need a greater rate of energy input, since some energy would be transferred as heat. The person is not 100% efficient at transferring chemical energy into work.

Example

A person of mass 60 kg runs up a flight of stairs that is 10 m high in 6 seconds. Find the power output of the person. (Take $g = 10$ N kg^{-1} to simplify the example.)

Answer

The person has to do work against gravity to lift their own weight.

force = $mg = 60 \times 10 = 600$ N

The distance moved in the direction of the force is 10 m. Work done:

$\Delta W = 600 \text{ N} \times 10 \text{ m} = 6000 \text{ J}$

If this work is done in 6 seconds, $\Delta t = 6$ s, $P = \frac{6000\text{ J}}{6\text{ s}} = 1000\text{ W}$ or 1 kW

Essential Notes

The force, F, and the velocity, v, must be in the same direction. If there was an angle of θ between the direction of the force and the velocity, the equation would become $P = Fv\cos\theta$

For a moving machine, such as a motor or a car, it is often useful to relate the power output to the velocity at which the machine is moving. We can write:

$$\text{power} = \frac{\text{work done}}{\text{time}} = \frac{\text{force} \times \text{distance}}{\text{time}}$$

This can be written as power = force $\times \frac{\text{distance}}{\text{time}}$

Since $\frac{\text{distance}}{\text{time}}$ = velocity,

power = force × velocity:

$$P = Fv$$

Efficiency

During an energy transfer, the total energy stays constant. However, the energy may not all be transferred as *useful* energy. For example, the engine of a car transfers chemical energy from the fuel to kinetic energy, but a significant amount of energy is transferred to thermal energy in the engine, tyres, road surface, etc. The proportion of the input energy that is transferred to useful energy is known as the **efficiency**.

Definition

$$\textit{Efficiency} = \frac{\textit{useful energy output}}{\textit{total energy input}}$$

Because energy = power × time, this can also be written

$$\textit{Efficiency} = \frac{\textit{useful output power}}{\textit{input power}}$$

Essential Notes

Efficiency can never be greater than 1. In practice it is always less. It is often expressed as a percentage. For example, the efficiency of a diesel engine is typically between 30% and 40%.

Example

An incandescent electric light bulb has an input power of 100 W but is only approximately 3% efficient. How much light energy does the bulb emit in one minute?

Answer

$$\text{Efficiency} = \frac{\text{useful output power}}{\text{input power}}$$

So

useful output power = input power × efficiency

In this case the useful power output is in the form of light and is equal to

100 W × 0.03 = 3 W or 3 joules per second

The total light energy output in one minute = 3 × 60 = 180 J

Essential Notes

Work is also done when a material is stretched or compressed, and energy is stored or released by the material. The amount of energy stored by a material can be determined using the area under a force–displacement graph (see Section 3.4.2.1).

3.4.1.8 Conservation of energy

When energy is transferred from one form to another, the total amount of energy does not change. This idea, known as the conservation of energy, is a fundamental principle in physics. It is not always obvious that the principle is obeyed. When a car brakes to halt at a junction it may appear that all the kinetic energy has disappeared, whereas in fact all the energy has been transferred to the brakes and the surroundings as thermal energy. The conservation of energy only applies to a *closed system*; the total energy will not stay constant if energy has been transferred to another object. In the example of the braking car we have to take into account the energy transferred in heating the air and the road.

Essential Notes

Since Einstein's work on relativity, the conservation of energy has had to be extended to include mass. In Section 3.2 you will have seen that energy can become mass in the process called pair production, where two particles are created from energy. Mass can also become energy in the process of annihilation when matter and antimatter meet and are converted to gamma radiation.

Definition

The ***principle of conservation of energy*** *states that the total energy of a closed system is constant.*

Kinetic energy

The energy that a moving mass has because of its motion is known as its **kinetic energy**. The kinetic energy of a moving mass depends on the mass and on the velocity squared.

Definition

Kinetic energy, $E_k = \frac{1}{2}mv^2$

The kinetic energy depends on velocity *squared*, rather than just velocity. So if a car doubles its speed, its kinetic energy will go up by a factor of four. This is why the stopping distance for a car goes up from 6 m at 20 mph to 24 m at 40 mph.

Example

Estimate the average braking force needed if a family car is to be stopped in 6 metres from a speed of 20 mph.

Answer

The mass of a typical family car is just over a tonne, say 1200 kg. Since there are 1.6 km in a mile, a speed of 20 mph is

$$\frac{1600 \times 20}{60 \times 60} = 8.9\,\text{m s}^{-1}$$

The kinetic energy would be:

$$E_k = \frac{1}{2}mv^2 = 0.5 \times 1200 \times (8.9)^2 = 47\,400\,\text{J}$$

This energy is used to do work against an average braking force F, so $E_k = F \times d$ or

$$F = \frac{E_k}{d} = \frac{47\,400}{6} = 7900\,\text{N}$$

Notes

A common mistake is to square the whole expression rather than just the velocity, e.g.

$$\left(\frac{1}{2} \times 1200 \times 8.9\right)^2$$

which would give 2.85 MJ, far too big an answer. Another common mistake is to forget to square the velocity.

Gravitational potential energy

Gravitational potential energy is the energy that an object has because of its position in a gravitational field. You need to do work to raise a mass to a greater height above the surface of the Earth, and this work is stored as potential energy. If the mass is allowed to fall, the potential energy will be transferred to the mass as kinetic energy.

The work done in lifting a mass, m, through a height, Δh, is:

$$\Delta W = \text{force} \times \text{distance} = (m \times g) \times \Delta h$$

This is also equal to the potential energy gained, ΔE_p, by the mass.

Definition

The change in the potential energy is the mass × gravitational field strength × the change in height,

$$\Delta E_p = mg\Delta h$$

The equation is only strictly true for places where the gravitational field strength, g, is constant. Although g does decrease as the distance from the Earth's surface increases, the equation is reasonably accurate for small values of Δh.

When a mass falls from a height, its potential energy is transferred to kinetic energy. If we can ignore energy losses due to air resistance, then all the potential energy will end up as kinetic energy: $\Delta E_p = \Delta E_k$.

Example

A high diving board is 10 m above the water surface. Calculate the speed at which a diver hits the water.

Answer

$\Delta E_p = \Delta E_k \qquad mg\Delta h = \frac{1}{2}mv^2$

So $v^2 = 2g\Delta h = 2 \times 9.81 \text{ N kg}^{-1} \times 10 \text{ m} = 196 \text{ m}^2 \text{ s}^{-2}$

$v = \sqrt{196} = 14 \text{ m s}^{-1}$

3.4.2 Materials

3.4.2.1 Bulk properties of solids

Density

The density of a material is the mass per unit volume.

$$\text{density} = \frac{\text{mass}}{\text{volume}} \quad \text{or} \quad \rho = \frac{m}{V}$$

In SI units, density is measured in kilograms per cubic metre, kg m^{-3}, though you will come across the use of grams per cubic centimetre, g cm^{-3}.

The densities of some materials are shown in Table 4. This section of the specification is about solids, but for comparison the density of water is 1000 kg m^{-3} and the density of air is 1.297 kg m^{-3} (at 0 °C and standard atmospheric pressure).

Essential Notes

You may need to convert a value given in g cm^{-3} to the correct SI units of kg m^{-3}. You should:

- Divide by 1000 (to convert grams to kilograms)
- Multiply by 1000 000 (to convert cm^{-3} to m^{-3})

The overall effect is to multiply by 1000. For example, pure water has a density of 1.00 g cm^{-3}, or 1000 kg m^{-3}.

Material	Density/ kg m^{-3}
aluminium	2700
bone	1700–2000
iron	7870
copper	8960
gold	17 650
mercury	13 690
water	1000
air (at 0 °C and standard atmospheric pressure)	1.297

Table 4
Densities of some common materials. You do not need to memorise these, but try to recall the order of magnitude of the values, e.g. $\rho_{air} \approx 1$ kg m^{-3}, $\rho_{water} \approx 1000$ kg m^{-3}, $\rho_{copper} \approx 10\,000$ kg m^{-3}

Example

What is the weight of water in a rectangular water tank that measures 0.80 m × 0.80 m × 0.50 m?

Answer

The volume of water is 0.32 m^3. The density of water is 1000 kg m^{-3}.

Since $\rho = m / V$,

$$m = \rho \times V$$

So the mass of water is 320 kg. Taking g to be 9.8 N kg^{-1}, the weight of water = mg = 3140 N (to 3 s.f.).

Materials under tension

When an object is subjected to opposing forces, it may be stretched or compressed (see Fig 36).

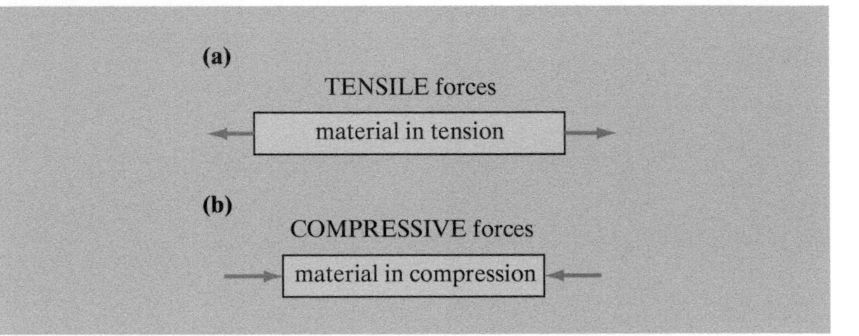

Fig 36
Showing the direction of the forces that act on objects in tension (a) and compression (b)

Tensile forces tend to stretch an object and cause an extension. How much the object stretches for a given force is often extremely important. For example, the cables carrying the roadway on a suspension bridge will stretch under the load, but by how much?

Springs are often important mechanical components in their own right, but the real importance to physicists is that the behaviour of a spring under tension can be a good model for the forces between atoms.

To investigate the behaviour of a spring under tension, the spring can be suspended vertically. Masses are then hung from the bottom of the spring and the extension is measured (Fig 37). Typical results from such an experiment are shown in Fig 38.

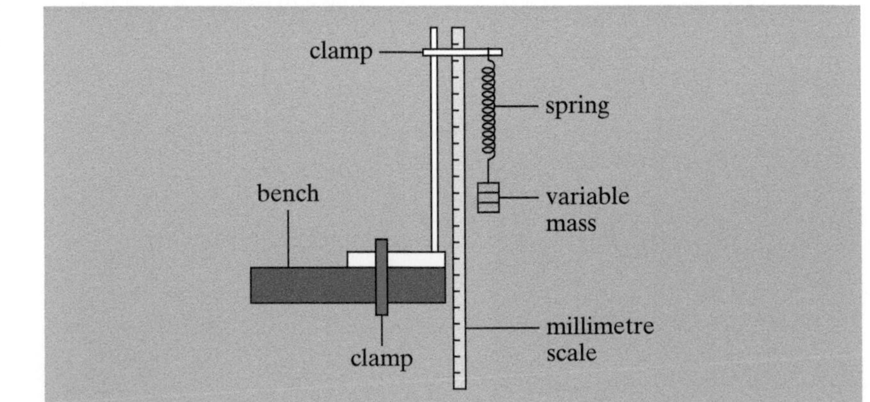

Fig 37
Stretching a spring

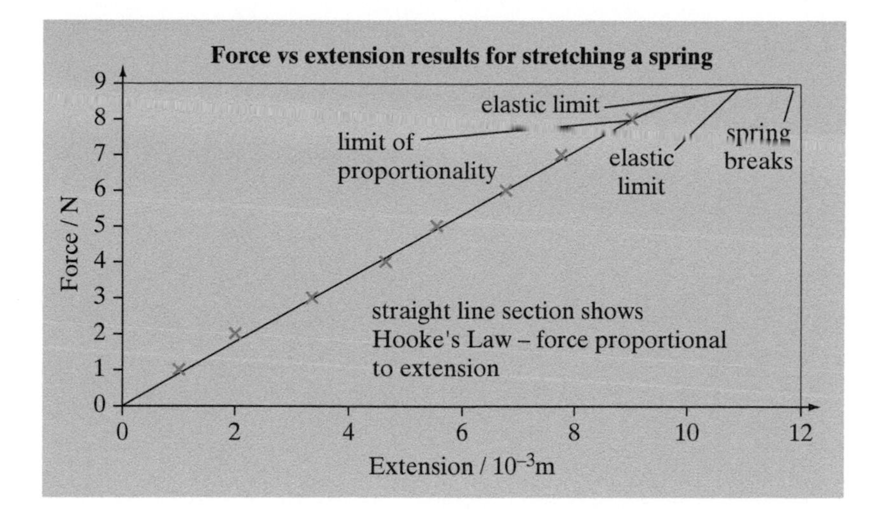

Fig 38
A force–extension graph for a spring under tension

The graph is a straight line (up to a point), which shows that the extension of a spring is proportional to the tensile force. This is known as **Hooke's Law** and can be written

$$F = k\,\Delta l$$

where Δl is the change in length of the spring (the extension), and k is the **spring constant**. The spring constant is a measure of the stiffness of the spring, i.e. how much force it takes to stretch it by a given distance. Its value is the gradient of the force–extension graph and its units are N m^{-1}.

Hooke's also applies to objects other than springs, for example wires. The law is not obeyed, however, if the force becomes too large – the object is then stretched beyond its limit of proportionality, and a relatively small increase in the force may cause a large extension. The extension may also be permanent; the object may remain deformed even when the force is removed.

Springs and wires tend to obey Hooke's Law at smaller loads, and also behave **elastically** up to a certain load, known as the **elastic limit**. They return to their original length when the tensile force is removed. However, if the force is large enough it may cause a permanent extension, so that the spring does not return to its original size, even when the force is removed. The spring is said to have been stretched past its **elastic limit**. Objects that show a permanent deformation, even when the force is removed, are said to show **plastic behaviour**.

Stress and strain

In order to compare the elastic properties of different materials, like copper , brass, iron, etc., it is important to take into account the dimensions of the sample. If the metals were in the form of wires for example, the cross-sectional area and the original length would affect how much they stretched for a given force (see Fig 39).

To allow for the effects of thicker wires, we define the property **stress** (or tensile stress), σ, as force per unit cross-sectional-area, A.

Definition

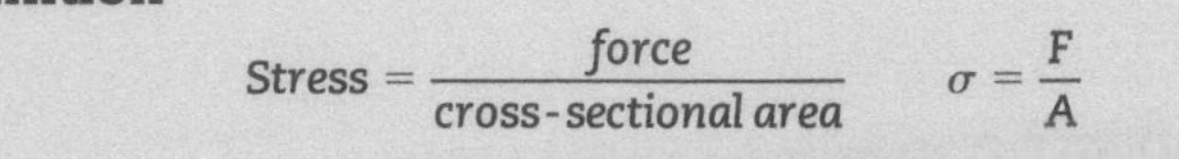

$$\text{Stress} = \frac{\text{force}}{\text{cross-sectional area}} \qquad \sigma = \frac{F}{A}$$

Stress is measured in pascals, Pa. 1 Pa is equivalent to 1 N m^{-2}.

Fig 39

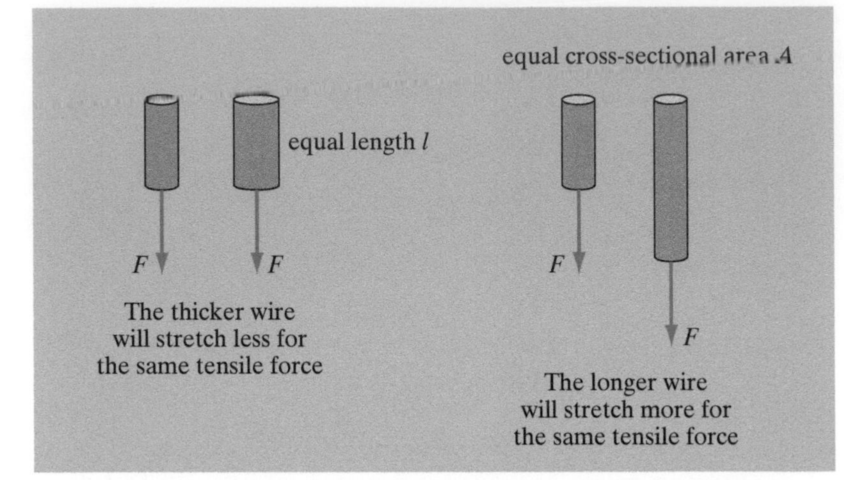

To allow for the effects of longer wires, we define the property **strain** (or tensile strain), ε, as the fractional extension, that is the extension, Δl, divided by the original length, l.

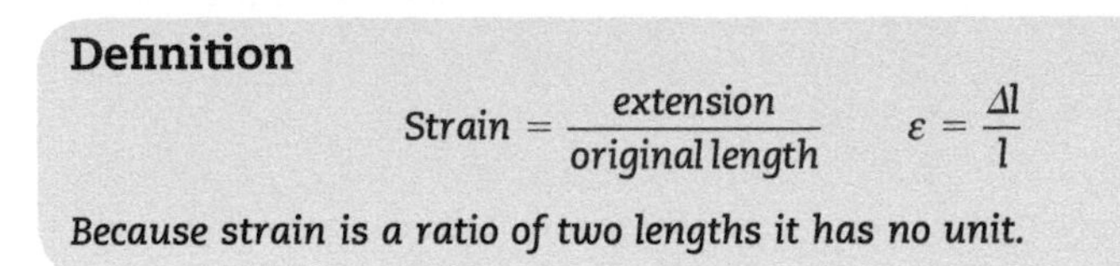

Definition

$$\text{Strain} = \frac{\text{extension}}{\text{original length}} \qquad \varepsilon = \frac{\Delta l}{l}$$

Because strain is a ratio of two lengths it has no unit.

The behaviour of materials under tension is investigated using a **tensile tester**, which automatically records the extension of a specimen whilst the load is increased. By plotting stress against strain, rather than force against extension, we can allow for the fact that specimens may have different dimensions and so get a direct comparison between different materials.

Materials such as copper wire, glass fibre and the rubber in rubber bands behave very differently under tension (see Fig 40).

Essential Notes

Force–extension graphs apply only to the *specimen* under test. Stress–strain graphs on the other hand apply to the *material* under test, regardless of the dimensions of the specimen.

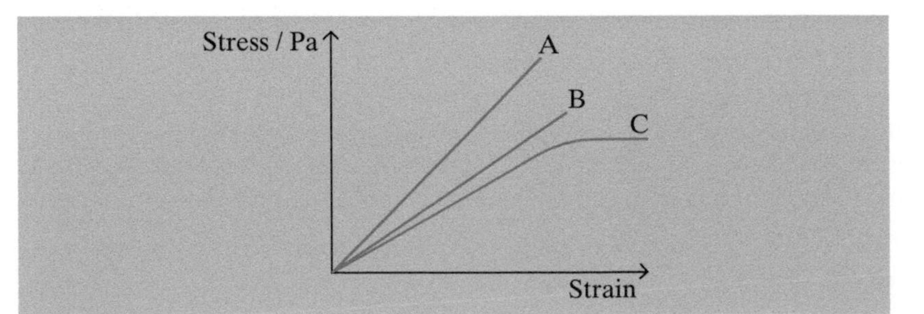

Fig 40
Stress–strain graphs for three different materials

We use a number of specialist terms to describe the behaviour of materials under tensile stress:

Elasticity	If a material returns to its original size and shape when the deforming force is removed, it is said to be behaving **elastically**.
Elastic limit	The maximum stress that can be applied to a material without causing a permanent extension.
Hooke's Law	The extension produced by a force in a wire or spring is directly proportional to the force applied. This only applies up to the limit of proportionality.
Yield point (stress)	The point at which a material continues to stretch even though no extra load is added.
Plasticity	Above the elastic limit, a material does not return to its original size and shape when the deforming force is removed. This is called plastic behaviour.
Strength	Strong materials are able to withstand large stresses before they fracture (break).
Breaking stress (or ultimate tensile stress)	The maximum stress that can be applied to a material without it breaking.
Stiffness	A measure of the force needed to change the size or shape of an object. The stiffness of a spring is measured in newtons per metre, N m^{-1}. The **Young modulus** is used as a measure of stiffness for different materials (see page 44). • Thick steel wire is stiffer than thin steel wire of the same length. • Short steel wire is stiffer than longer steel wire of the same diameter. • Steel is stiffer than copper of the same diameter and length, because the copper extends more per unit force.
Ductility	A ductile material can undergo large plastic deformation before it fractures. Copper is a ductile material that can be drawn under tension into wire.
Brittleness	A brittle material cannot be plastically deformed. The material fractures when the yield stress is exceeded. Brittle materials, such as cast iron and concrete, tend to be strong in **compression** but fracture easily under tension.

Example

Study the stress–strain graphs in Fig 40.

(a) Which material is the stiffest?

(b) Which material is the most ductile?

(c) Which material is the strongest?

Answer

(a) A produces the steepest gradient and is therefore the stiffest.

(b) C has the longest plastic region and is therefore the most ductile.

(c) A has the highest breaking stress and is therefore the strongest.

3.4.2.2 The Young modulus

Definition

The **Young modulus** *(E) is a measure of the stiffness of a material, and is given by:*

$$E = \frac{\text{tensile stress}}{\text{tensile strain}}$$

The Young modulus has units of N m^{-2}, *or pascal,* Pa.

If F is the tensile force, A is the cross-sectional area, Δl is the extension and l is the original length,

$$E = \frac{F}{A} \div \frac{\Delta l}{l} = \frac{F}{A} \times \frac{l}{\Delta l} \text{ or } \frac{Fl}{A\Delta l}$$

The Young modulus can be used to compare the stiffness of different materials, even if the samples under test have different dimensions.

The Young modulus applies only up to the limit of proportionality.

Example

A 1.50 m long steel piano wire, with a diameter of 1.00 mm, is stretched by a force of 50.0 N.

(i) Calculate the stress in the wire.

(ii) Calculate the increase in length of the wire.

(iii) Calculate the energy stored in the wire.

(Young modulus of steel = 210×10^9 Pa)

Answer

(i) Cross-sectional area of wire $= \pi\left(\frac{d}{2}\right)^2 = \pi \times (0.50 \times 10^{-3})^2$

$= 7.85 \times 10^{-7}\ \text{m}^2$

$$\text{Stress} = \frac{\text{force}}{\text{area}} = \frac{50.0}{7.85 \times 10^{-7}}\ \text{N m}^{-2} = 6.37 \times 10^7\ \text{Pa} \approx 64\ \text{MPa}$$

(ii) Strain $= \dfrac{\text{stress}}{E} = \dfrac{6.37 \times 10^7}{210 \times 10^9} = 3.03 \times 10^{-4}$ (no units)

Extension $\Delta l = l \times \text{strain} = 1.5 \times 3.03 \times 10^{-4} = 4.55 \times 10^{-4}\,\text{m} \approx 0.50\,\text{mm}$

(iii) Energy $= \frac{1}{2}F\Delta l = 0.5 \times 50.0 \times 4.55 \times 10^{-4} = 1.1 \times 10^{-2}\,\text{J}$

Experimental determination of the Young modulus

The Young modulus of a material in the form of a wire can be determined using the apparatus shown in Fig 41.

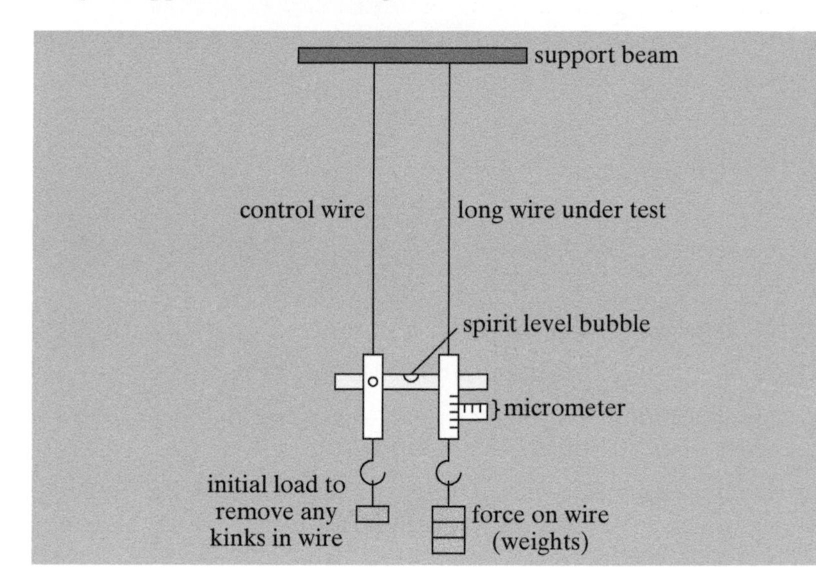

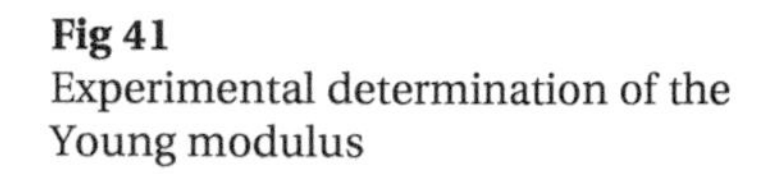

Fig 41
Experimental determination of the Young modulus

Essential Notes

This is often referred to as Searle's apparatus.

The wire, which should be at least one metre long, is suspended vertically from a strong support and stretched by hanging weights on the lower end (see Fig 41). The key points of the experimental method are:

- a metre rule is used to measure the original length, *l*, of the test wire
- a micrometer is used to measure the diameter of the wire; several readings are taken at different points along the wire to account for variations in the wire's thickness (the cross-section of the wire may not be perfectly circular so it is a good idea to take measurements of the diameter at different angles)
- the mean diameter, and hence radius, is calculated and used to find the cross-sectional area in m^2
- the wire is suspended in parallel with a control wire (as in Fig 41)
- both wires are initially loaded to straighten them
- the micrometer attached to the test wire is adjusted in order to bring the spirit level horizontal
- the micrometer reading is then taken
- the load on the test wire is increased, the micrometer is adjusted to re-level the apparatus and a new reading is taken
- this is repeated to obtain a set of results showing how the load on the wire affects its extension

- a second set of measurements is taken as the weights are removed
- the readings are used to calculate stress and strain values
- a graph of stress (y-axis) against strain (x-axis) is plotted.

Analysing the results of the Young modulus experiment

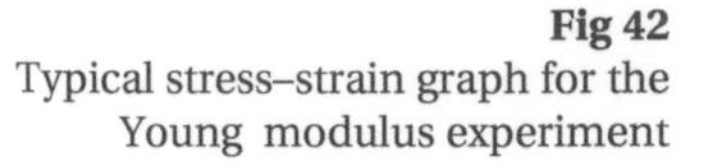

Fig 42
Typical stress–strain graph for the Young modulus experiment

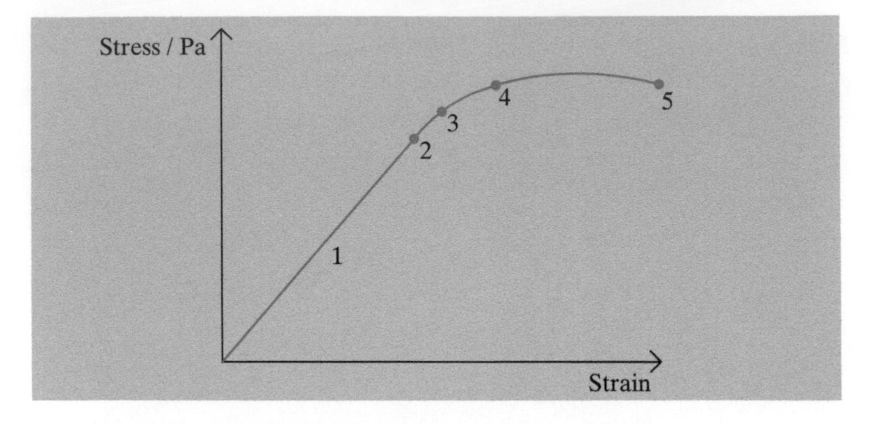

In Fig 42 the following regions and points are marked:

1 Linear region – extension ∝ force (obeys Hooke's Law).

2 Proportionality limit – beyond this point extension is no longer proportional to force (limit of Hooke's Law).

3 Elastic limit – material begins to behave plastically. This is the point beyond which, when the stress is removed, the material does not return to its original length.

4 Yield point – material shows large increase in strain for small or no increase in stress.

5 Breaking stress (or ultimate tensile stress) – the applied stress causes the material to fracture (break).

The behaviour of ductile and brittle materials is different when they are stretched to breaking point.

Ductile materials

When a ductile material (e.g. copper wire) is subjected to a high tensile stress it undergoes considerable plastic deformation (Fig 43). During this plastic stage, and just as it is about to fail, the material **necks** (Fig 44).

Fig 43
Stress–strain graph of a ductile material

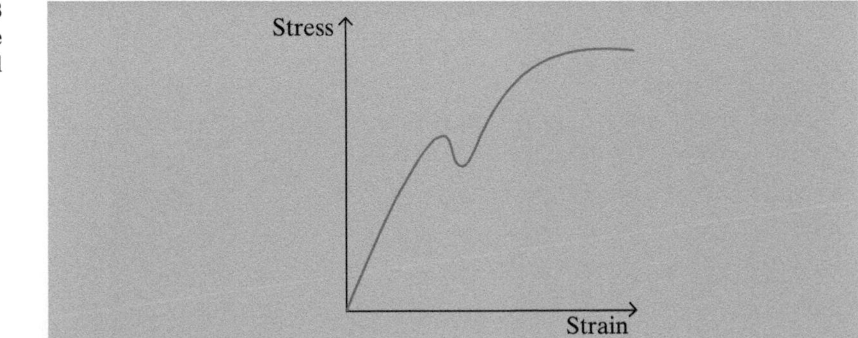

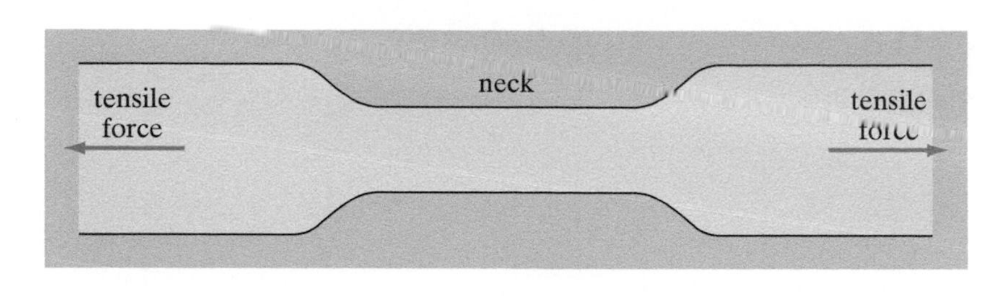

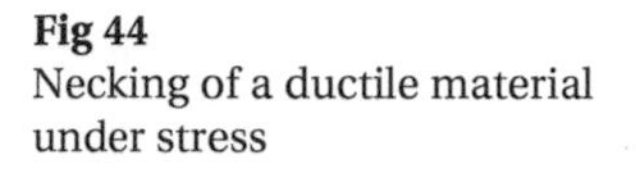
Fig 44
Necking of a ductile material under stress

The cross-sectional area of the wire gets narrower, and because stress = force/area, the neck experiences an increase in tensile stress. The material fails at the neck. The neck provides an early warning that the material is about to fail.

Brittle materials

Brittle fracture is due to the rapid extension of surface or internal cracks in the material. The fracture is sudden and catastrophic. There is little or no plastic deformation as a warning (Fig 45). There is a concentration of stress around the cracks; the sharper the crack, the greater the stress. Crack movement is easier under tension and more difficult under compression. Pre-stressing brittle materials can increase their strength.

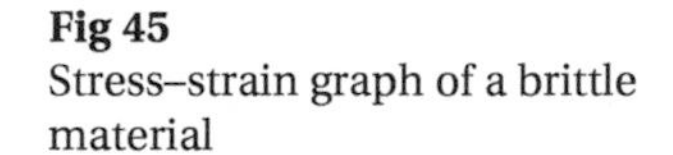
Fig 45
Stress–strain graph of a brittle material

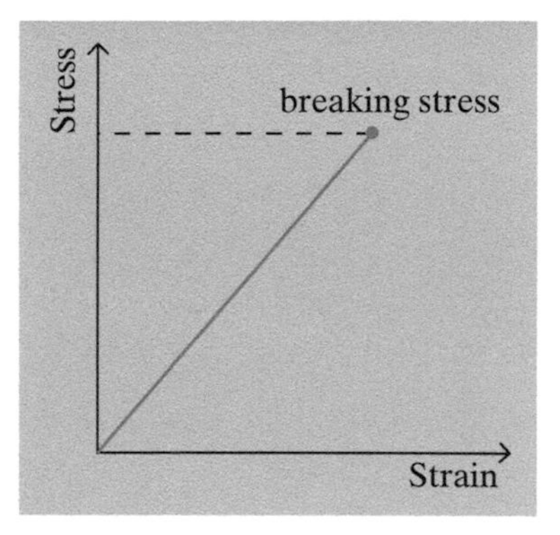

Elastic strain energy

When a wire is stretched by a force, provided the elastic limit is not exceeded, then the work done (energy change) is stored as elastic potential energy, or **elastic strain energy**, in the wire. The area below the graph line is the total work done in stretching the wire, and is therefore equal to the elastic strain energy stored. This can be shown as follows.

Word done by force F in extending the wire by a small extension δl is

$$\delta W = F\,\delta l$$

Total work done in fully extending the wire is therefore the sum of the small areas $F\,\delta l$ (see Fig 46).

$$W = \sum \delta W = \sum F\,\delta l$$
$$= \text{area under graph}$$

Area of triangle $= \frac{1}{2} \times \text{base} \times \text{height} = \frac{1}{2} \times F \times \Delta l$

$$\text{Elastic strain energy} = \tfrac{1}{2} F \Delta l$$

Elastic strain energy is more obvious in a spring where the energy stored during stretching is released again to restore the spring to its original shape.

Since $F = k\,\Delta l$ for a spring (see page 32)

$$\text{Elastic strain energy stored in a spring} = \tfrac{1}{2} k (\Delta l)^2$$

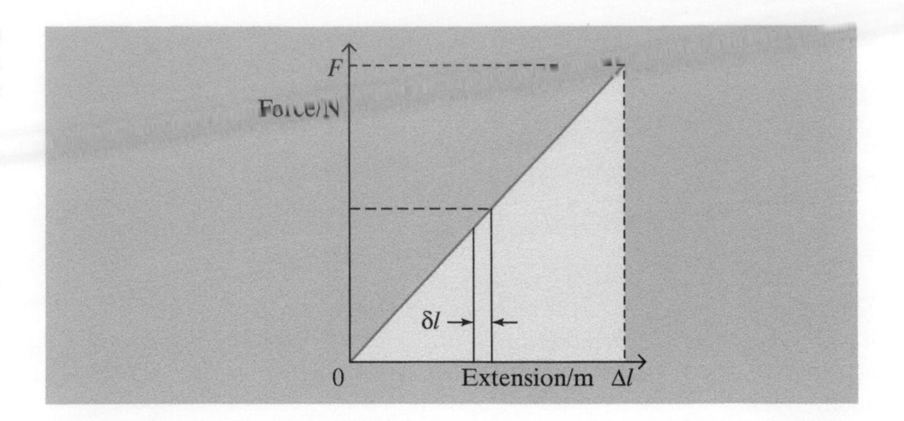

Fig 46
Calculating the work done in stretching a wire

Example

Therabands are strips of rubber that can be stretched by people who are exercising. They are particularly useful in physiotherapy to help patients recover after surgery. Therabands come in different stiffnesses. The graph below shows the force–distance graph as the Theraband is stretched (blue) and gradually released (red).

(a) Explain what is meant by 'stiffness', and state what units it should be measured in.

(b) The stiffness of the Theraband used for the graph was advertised as 600. Is this a fair description? Explain your answer.

(c) The variation of force with distance (extension) depends upon whether the strip is being stretched or released. Explain what this means in terms of energy transfer for this Theraband.

(d) Estimate the work done in stretching the band by 0.75 m.

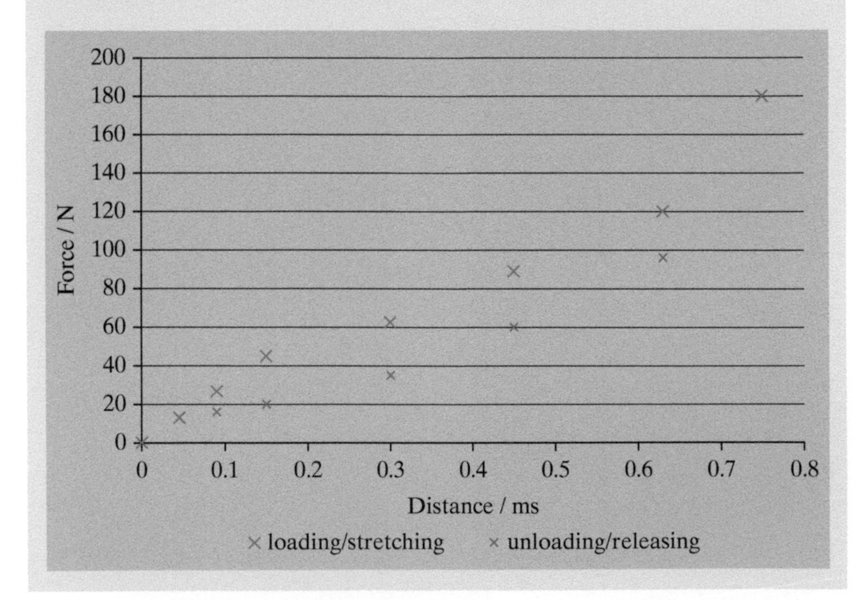

Answers

(a) Stiffness measures how difficult the band is to stretch, i.e. the force required to change the length. It is measured in Newtons per metre (N m^{-1}).

(b) Stiffness is calculated from the slope of the graph. The central part of the graph from approximately 0.2 to 0.6 is fairly straight and has a value of approximately 600 N m^{-1}. However, the Theraband does not follow Hooke's Law across the full range of results and is considerably stiffer for the first 0.1 m and between 0.6 and 0.8.

(c) The energy transferred (or work done) in stretching the Theraband is equal to the area below the curve. The area beneath the blue curve is greater than the area beneath the red curve. This means that more work is done in stretching the Theraband than is transferred back when the Theraband is released. The 'missing' energy has been dissipated internally as heat.

(d) To estimate the area below the curve, break up the graph into sections. The area lies approximately between 65 J and 75 J.

The energy stored in a spring can be transferred to kinetic energy, as in a clockwork toy. The springs in a car suspension system are compressed by the mass of the car and its load. They may be further compressed by bumps in the road, as the car is driven. The work done in compressing the springs is then returned to the car as kinetic energy and gravitational potential energy (lifting the car!). This continual transfer of energy would make the car bounce up and down, making it impossible to drive safely. Shock absorbers are fitted to prevent this happening. Shock absorbers (or dampers) consist of a piston moving in a cylinder, similar in design to a bicycle pump. However, in the shock absorber the cylinder is full of oil. As the piston moves through the oil, it transfers energy as heat to the oil, reducing the kinetic energy, which in turn reduces the amount by which the car bounces.

Example

A pogo stick has a spring which is compressed by 2 cm when Ben, who has a mass of 70 kg, stands on it.

(a) Calculate the spring constant.

Ben jumps so as to compress the spring by a further 3 cm.

(b) Calculate the energy stored in the spring.

(c) If all of the elastic potential energy stored in the spring is transferred as gravitational potential energy, how high would Ben bounce?

Answers

(a) $k = F / x = (70 \times 9.81) / 0.20 = 3.4 \times 10^4$ N m^{-1}

(b) $E = ½\ kx^2 = ½ \times 3.4 \times 10^4 \times (0.05^2) = 43$ J

(c) $mgh = 43$ J, so $h = 6.25$ cm

The Young modulus is the gradient of the stress-strain graph.

The precision and accuracy of the experiment are improved by the following steps:

- a long thin wire is used to produce a large extension without the need for an excessively high load
- a control wire is used so that changes in length due to temperature changes, or to a sagging support, do not affect the results
- a metre rule is **accurate** enough for measuring the length of the wire since an error of a few mm in 1 m is not significant. $\left(\text{e.g. } \frac{2}{1000} \times 100 = 0.2\%\right)$
- a micrometer with a resolution of 0.01 mm is needed to measure the diameter and extension of the wire. For a wire of 0.5 mm diameter, this gives an error of $0.01/0.5 \times 100 = 2\%$ in the diameter. This is a *significant* error, because it leads to an error of $2 \times 2 = 4\%$ in the cross-sectional area, A.

Notes

The Young modulus is the gradient of the linear section of a stress–strain graph. Choose a large part of this section for your calculation of gradient as this improves the precision of your answer.

Example

A material in the form of a wire, 1.0 m long, is subjected to a tensile force and a stress–strain graph of loading and unloading the wire is drawn. The graph is shown in Fig 47.

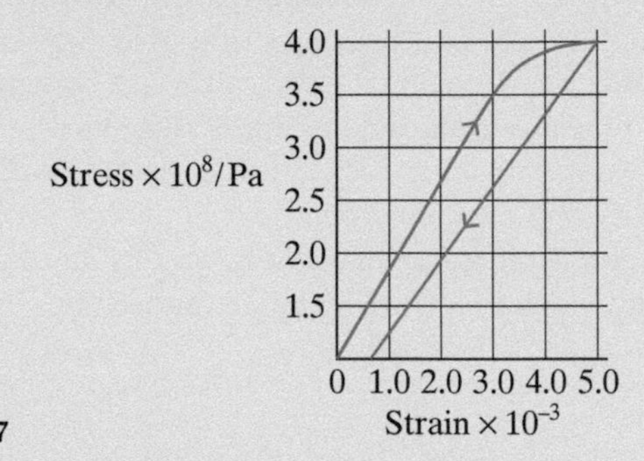

Fig 47

(i) Explain the shape of the graph.

(ii) Use the graph to determine at what extension the limit of proportionality occurs.

(iii) Use an appropriate part of the graph to determine the Young modulus for the material.

(iv) Use the information in Table 5 to determine which material the wire is made of.

Material	Young modulus/GPa
aluminium	71
brass	100
copper	117
gold	71
iron	206
silver	70
stainless steel	200
zinc	110

Table 5

Answer

(i) For stresses of up to 3.5×10^8 Pa (the limit of proportionality) the wire obeys Hooke's Law (extension $\propto$ force). Beyond this point the elastic limit is reached. The material undergoes plastic deformation. This indicates that the material is ductile. When the stress is removed the material has undergone a permanent change in length. The graph does not return through the origin.

(ii) Strain $= \frac{\Delta l}{l}$ so extension $\Delta l = \text{strain} \times l = 3.0 \times 10^{-3} \times 1.0$

$= 3.0 \times 10^{-3}$ m or 3 mm

(iii) $E = \frac{\text{stress}}{\text{strain}} = \frac{3.5 \times 10^8}{3.0 \times 10^{-3}} = 1.17 \times 10^{11}$ Pa

(iv) Table 5 reveals that copper has a Young modulus of 117×10^9 Pa.

3.5 Electricity

3.5.1 Current electricity

3.5.1.1 Basics of electricity

Matter is composed of atoms, which contain positively charged protons and negatively charged **electrons**. These oppositely charged particles attract each other, holding atoms, molecules and matter itself together. These positive and negative charges can be separated, perhaps by friction as when a balloon is charged by rubbing it against your clothes or chemically as in a battery. The separation of positive and negative charges creates a **potential difference**, which can lead to electrostatic phenomena, such as a balloon sticking to the wall, a small electric shock when you touch a car door or a flash of lightning. When the charged particles move, as in a spark, the flow of charge is known as an **electric current**.

Definition

An electric current is the rate at which electric charge flows past a given point.

A current of 1 A = a charge flow of 1 C per second, which can be written as:

$$I = \frac{\Delta Q}{\Delta t}$$

where I = current in amperes (A), ΔQ = change in charge in coulombs (C) and Δt = time interval in seconds (s)

Essential Notes

The ampere is actually defined in terms of the magnetic effect of an electric current (see AQA Physics A-level Year 2), and the coulomb is then defined in terms of the ampere.

Electric charge, Q, is measured in units of coulombs, C. The charge carried by an electron is small, equal to 1.6×10^{-19} C, which means that 1 C of charge is approximately 6×10^{18} electrons. In SI units electric current, I, is measured in amperes, A. A current of 1 ampere means that one coulomb of charge flows past a given point every second.

Essential Notes

Electrons carry a negative charge, so they flow in the opposite direction to conventional current. This can cause confusion, so be careful: the arrows showing the direction of flow of the current point the opposite way to the movement of electrons.

An electric current can be due to a flow of positive or negative charges, depending on the material that it is flowing through. In a plasma (a gas containing ions), or an electrolyte or a semiconductor, the charge carriers can be positive or negative or both. By convention the direction of electric current is taken as the direction that positive charge moves in.

In a metallic conductor, the charges are carried by moving electrons, so a current of 1 A means that approximately 6×10^{18} electrons are flowing past in one second.

Potential difference

A potential difference (abbreviated to p.d.) is needed to make an electric current flow. The potential difference can be produced by many different devices: a dry cell, a battery, a dynamo or generator, a photovoltaic cell, etc. All of these devices transfer energy to electric charges. The larger the amount of energy transferred to each charge, the higher the potential difference. As the electric charges flow, around an electrical circuit for example, energy will be transferred from the charges, as heat in wires, light in lamps, motion in motors etc.

Essential Notes

Since $V = \frac{W}{Q}$, the work done (or energy transferred) by a charge Q, moving through a potential difference V is given by

$$W = QV$$

This may sometimes be written as E (energy) = QV.

For example, in an X-ray tube an electron may be accelerated through a p.d. of 100 kV. Since $E = QV$, the electron would gain an energy of

$$1.6 \times 10^{-19}\ \text{C} \times 100\,000\ \text{V} = 1.6 \times 10^{-14}\ \text{J}.$$

Definition

The potential difference between two points is defined as the electrical energy transferred to, or from, a unit charge as it passes between the points. This can be expressed as:

$$\text{potential difference} = \frac{\text{energy transferred (or work done)}}{\text{charge}} \quad \text{or} \quad V = \frac{W}{Q}$$

In SI units, V is measured in volts, W is measured in joules and Q is measured in coulombs.

So 1 volt = 1 joule per coulomb, $1\ \text{V} = 1\ \text{J C}^{-1}$.

Current, potential difference and resistance

A potential difference causes charges to move through a material, a metal wire for example. For a given potential difference, the size of the current that flows depends on certain properties of the material:

- How many charge carriers are free to move? (In the case of a metal wire, how many free electrons are there?)
- How easy is it for the charges to move through the material? As charges flow through the material they interact with its atoms and transfer energy to the material, reducing the speed of the charges.

A material's opposition to current flow is called its **resistance**, R.
It is a measure of the potential difference, V, required to create a certain current, I.

Definition

Resistance is defined by the equation

$$R = \frac{V}{I}$$

In SI units, the potential difference is measured in volts and the current in amperes. The SI unit of resistance is the ohm, Ω.

When a potential difference of 1 volt across a component causes a current of 1 ampere to flow, the component has a resistance of 1 Ω.

1 Ω = 1 volt per amp

Prefix	Symbol	Multiplier	Current	Potential difference	Resistance
micro-	µ	$\times 10^{-6}$ A	µA, 10 µA : smallest potentially fatal current	µV, ouput from an EEG.	µΩ, a cubic centimetre of silver.
milli-	m	$\times 10^{-3}$ A	mA, 20 mA current through LED.	mV, Action potential from a nerve cell	mΩ, 1 km of overhead power line.
kilo-	k	$\times 10^{+3}$ A	kA, Current from transformer at electricity substation.	kV, Taser	kΩ, human body
Mega-	M	$\times 10^{+6}$ A	MA Plasma current in fusion experiments	MV National Grid power lines.	MΩ, ceramic insulators on power lines

Table 6
Prefixes for SI units of electrical quantities. The examples quoted are typical, order of magnitude values.

Notes

It is important to use the correct units. Table 6 will help you to convert other units to volts, amps and ohms. A common error is to confuse milliamps and microamps.

Example

In a conductor the charge carriers each have a charge of 1.6×10^{-19} C.

(a) Calculate the number of charge carriers passing a point in the conductor per second if the current is 4.0 μA.

(b) Calculate the p.d. generated by the charges across a 1000 ohm resistor.

(c) Calculate the work done per charge carrier.

Answer

(a) $Q = It = 4.0 \times 10^{-6} \times 1$

$= 4.0 \times 10^{-6}$ C

$$\text{Number of charge carriers} = \frac{\text{total charge}}{\text{charge on charge carrier}}$$

$$= \frac{4.0 \times 10^{-6}\ \text{C}}{1.6 \times 10^{-19}\ \text{C}}$$

$= 2.5 \times 10^{13}$ (This value has no units.)

(b) $V = IR = 4.0 \times 10^{-6} \times 1000$

$= 4.0 \times 10^{-3}$ V (or 4 mV)

(c) $W = VQ = 4.0 \times 10^{-3} \times 1.6 \times 10^{-19}$

$= 6.4 \times 10^{-22}$ J

3.5.1.2 Current–voltage characteristics

If we increase the potential difference across a sample of a material, like copper, or a component, such as a light bulb, we will cause a larger current to flow. The relationship between the potential difference, V, across a component, and the current, I, that flows through it can be found by experiment.

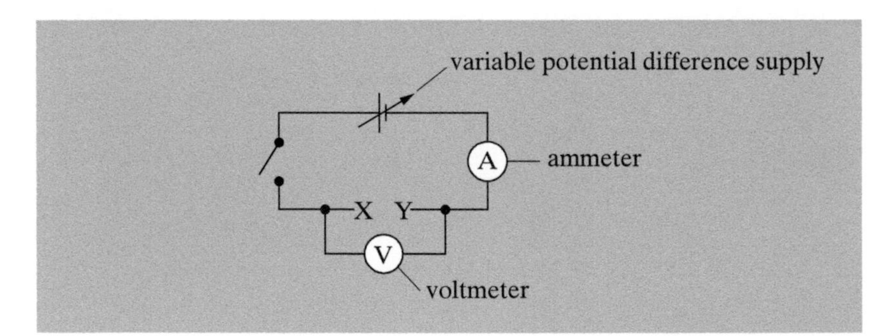

Fig 48
Circuit used to produce current/voltage characteristics of components

Essential Notes

We normally assume that an ammeter has zero resistance and a voltmeter has infinite resistance

A potential difference is applied to the component. A voltmeter is used to measure this potential difference. The voltmeter is connected across the component, i.e. it is in parallel with the component, since it is measuring a difference between two points (see Fig 48). Voltmeters have to have a very high electrical resistance, so that only a small current flows through them.

The current flowing through the component is measured using an ammeter. The ammeter is connected in series with the component, so that the same current flows through both. An ammeter has to have a very small resistance, or it would reduce the current it was being used to measure.

- The component under test is connected between points X and Y.
- The potential difference and current are measured.
- The potential difference of the supply is varied and the meters are read again.
- The polarity of the supply can be reversed, so as to investigate current when a reverse potential is applied.
- These current and potential difference values are used to plot a graph, known as the *I*–*V* characteristic.

There are three components that you need to study at AS or A-level. The results and conclusions for these are outlined below.

1 A resistor (for example made of carbon film or a metal wire)

When the current is plotted against the p.d., a straight line graph is obtained (Fig 49).

The results show that for a resistor, over the range of p.d.s investigated, the current that flows is proportional to the applied potential difference.

V I, or $V = IR$, where R is a constant (in this case)

Components which follow this rule are known as **Ohmic conductors.**

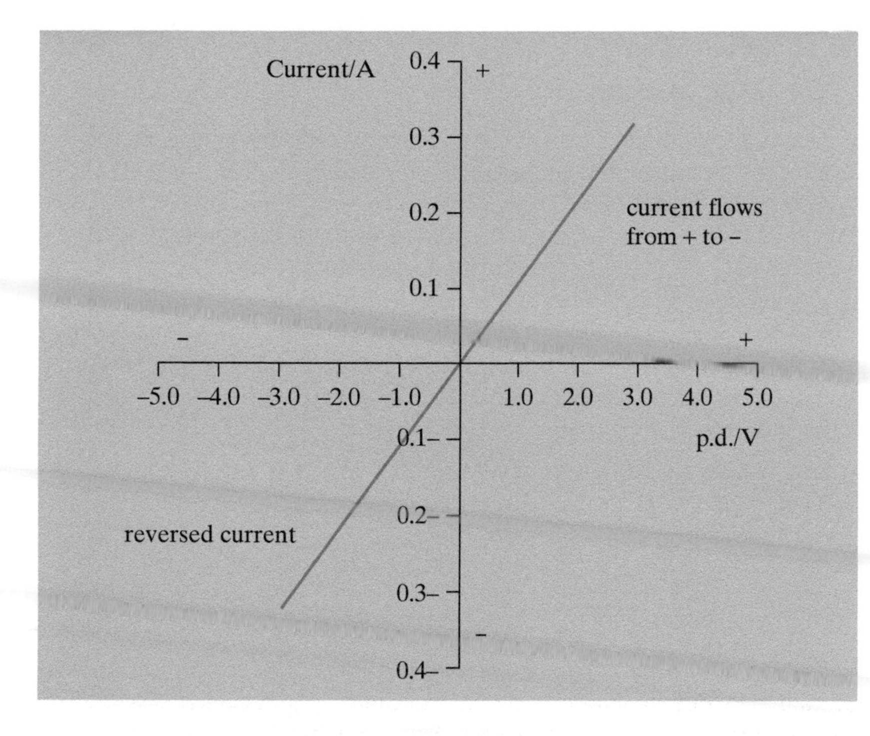

Fig 49
Current plotted against p.d. for an ohmic conductor

Since the gradient = I/V, the resistance can be found by calculating 1/gradient.

Reversing the polarity of the supply simply reverses the direction of the current. But the linear relation breaks down if the current is too large, since that causes a heating effect and electrical resistance is affected by temperature.

2 The filament of an incandescent light bulb

The filament of an incandescent bulb is a metal wire. But, unlike case 1 (above) the wire gets very hot. As the temperature of the wire increases, the moving charge carriers (electrons) lose more energy in their collisions with atoms in the wire, so there is an increase in resistance to current.

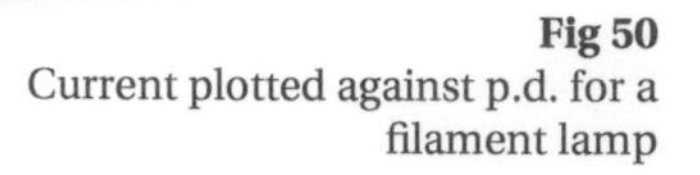
Fig 50
Current plotted against p.d. for a filament lamp

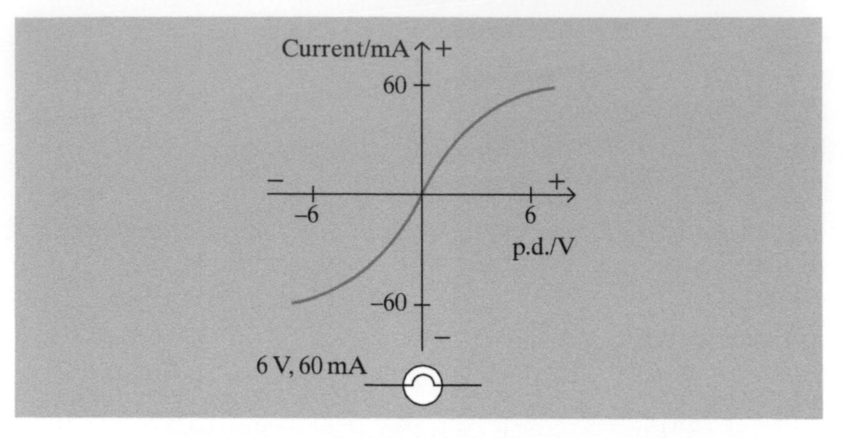

The resistance, R, at any given potential difference, V, can be found from this graph, since R is defined as $R = V/I$. However, the resistance is not constant, V is not proportional to R and the filament is described as a **non-ohmic conductor.**

3 A semiconductor diode

Essential Notes

The **diode** is a one-way device. It acts like a valve.

A semiconductor diode differs from the filament bulb and the resistor because the polarity of the potential difference **does** have an effect. The diode only conducts when the p.d. across it is approximately +0.7 V. When that happens, the diode is said to be **forward biased** and has a low resistance. If the p.d. across the diode is less than +0.7 V, the diode does not conduct; it has a very high resistance. If the potential difference across the diode is reversed, only a small 'leakage' current flows unless the potential is greater than the breakdown potential for the diode (between 50 and 500 V). In that case a large current flows, causing a heating effect that usually breaks the diode.

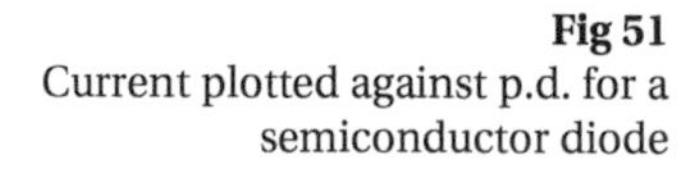
Fig 51
Current plotted against p.d. for a semiconductor diode

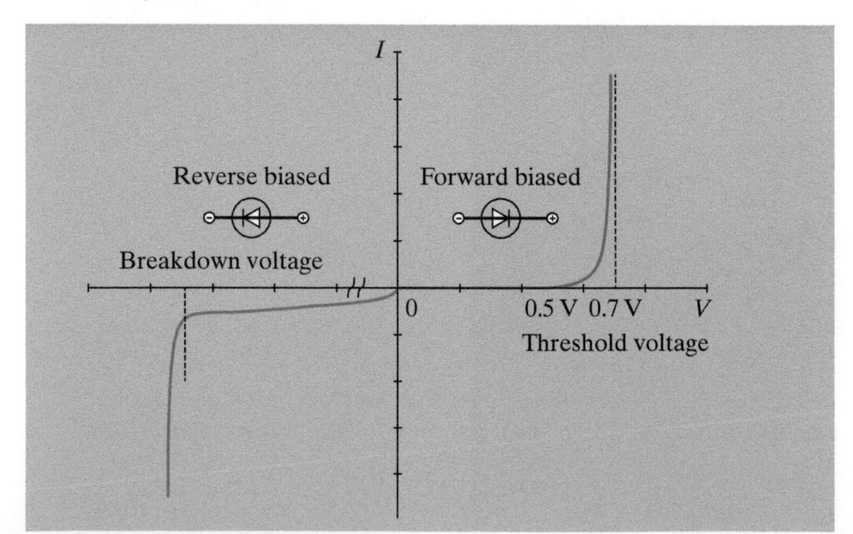

Ohm's Law

Ohm's Law is a special case and only applies to certain components in certain conditions.

Definition

***Ohm's Law** states that the current in a conductor is directly proportional to the p.d. across it,*

$$I \propto V$$

provided that the temperature and other physical conditions remain the same.

Essential Notes

The equation

$$R = \frac{V}{I}$$

defines resistance. This equation can always be used to calculate R for any conductor when a particular current flows.

The current/voltage characteristics on pp 56–58 can show clearly whether or not a component obeys Ohm's Law. The graphs in Figs 49, 50 and 51 show that the resistor/wire obeys Ohm's Law, while the semiconductor diode and filament lamp do not.

3.5.1.3 Resistivity

The resistance, R, of a sample of conducting material depends on:

- Its cross-sectional area, A. This is the area perpendicular to the direction of current flow. A larger cross-sectional area allows a larger current to flow; in fact resistance is inversely proportional to cross-sectional area:

  $$R \propto \frac{1}{A}$$

  Doubling the cross-sectional area will halve the resistance.

- Its length, l. A longer sample means that charge carriers, electrons in a metal wire for example, will have more interactions with atoms in the sample. A higher potential difference would be needed to provide the energy for the same current. In other words, the resistance increases.

  Resistance is proportional to length:

  $$R \propto l$$

- The number and mobility of charge carriers, for example electrons, in the material.

These factors determine the **resistivity**, ρ, of the material.

Notes

Doubling the radius or diameter of a wire would increase the area by a factor of 4, since

area $\propto$ radius^2

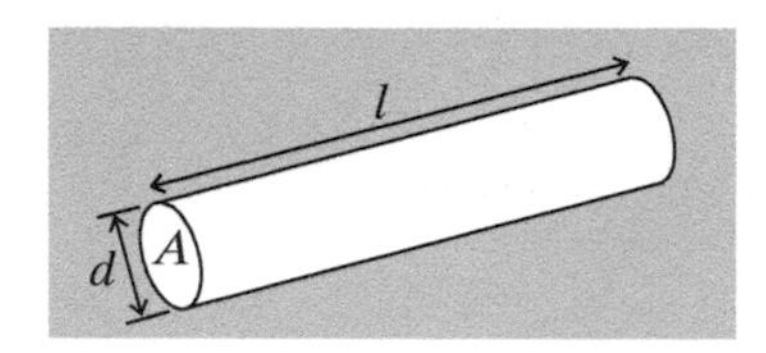

Fig 52

Definition

*The **resistivity** ρ of a material is given by:*

$$\rho = \frac{AR}{l}$$

This equation can be rearranged to give:

$$R = \frac{\rho l}{A}$$

where R is the resistance of the conductor, l is its length and A is its cross-sectional area. The resistivity ρ is a constant of the material from which the conductor is made and is measured in ohm metres (Ω m).

Notes

The unit of resistivity is the ohm metre (Ω m). A common error in exams is to express it as Ω m^{-1}.

Example

The graph shows how the potential difference across a wire X affects the current through it.

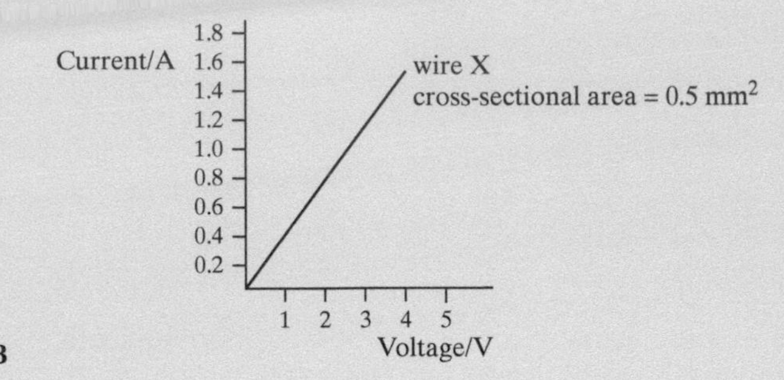

Fig 53

(a) Calculate the resistance of wire X.

(b) Calculate the resistivity of wire X if its length is 1.5 metres.

(c) Y is made from the same material as X. The resistance of 1.5 metres of Y is half that of X for the same length. Calculate the diameter of Y.

Answer

(a) Measure the inverse gradient of the graph, e.g. $R = \frac{V}{I} = \frac{3}{1.1} = 2.73\ \Omega$

(b) Rearrange $R = \frac{\rho l}{A}$

$$\rho = \frac{RA}{l} = \frac{2.73 \times 0.5 \times 10^{-6}}{1.5} = 9.1 \times 10^{-7}\ \Omega\ \text{m}$$

(c) R for Y is $\frac{2.73}{2} = 1.37\ \Omega$

Rearrange $R = \frac{\rho l}{A}$

$$A = \frac{\rho l}{R} = \frac{9.1 \times 10^{-7} \times 1.5}{1.37} = 9.96 \times 10^{-7}\ \text{m}^2$$

$$A = \pi\left(\frac{d}{2}\right)^2$$

Rearranging, $d = 2\sqrt{\frac{A}{\pi}} = 1.1 \times 10^{-3}$ m or 1.1 mm

Notes

Tolerance is usually applied when reading values from a graph, appropriate to the scale given.

Notes

Be careful with significant figures and quote answers to an appropriate value indicated by the data given in the question. A common error is to quote answers to one or two decimal places – this is not the same as significant figures.

Notes

Be careful when converting mm² to m²; divide by 10^6 and not 10^3.

Measuring resistivity

We can examine the factors that affect resistance by experiment, using the apparatus shown in Fig 54.

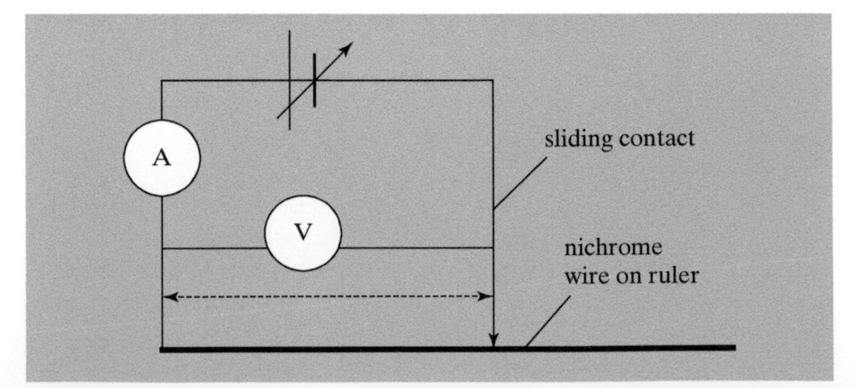

Fig 54
Measuring the resistivity of a material in the shape of a wire

By varying the length of the wire under test, we can find an accurate value for the resistivity of the metal.

- Start by measuring 100 cm of the wire under test. Tape the wire on to a metre rule, to avoid any kinks or twists. Connect the wire to the circuit using crocodile clips.
- Record, in a table, the p.d. displayed on the voltmeter and the current displayed on the ammeter, for this length of wire.
- Move the voltmeter connection along the wire in the range 100 cm to 30 cm and record the p.d. and current for each length.
- Calculate the resistance of wire for each recorded length using $R = \frac{V}{I}$.
- Measure the diameter of the wire, several times over its length, using a micrometer, to obtain a value for the mean diameter.
- Use the mean diameter to calculate the cross-sectional area using

  $$A = \pi\left(\frac{d}{2}\right)^2$$

- Since

  $$R = \frac{\rho l}{A}$$

  then R is proportional to length, for a wire of a given, constant cross-sectional area.
- Plot a graph of R on the y-axis against l on the x-axis, which should produce a straight line of gradient ρ/A.

Essential Notes

Avoid large currents which will heat the wire and increase the resistance.

Essential Notes

A multimeter set on the ohms range could be used to measure resistance directly, instead of using a battery, ammeter and voltmeter. However, the ohms range usually has an uncertainty of $\pm 1\ \Omega$.

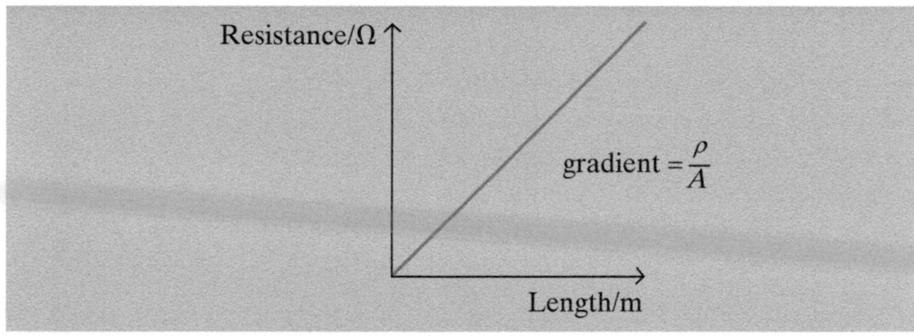

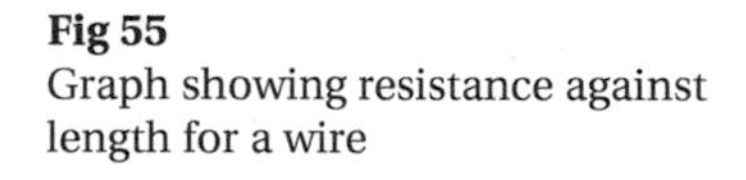

Fig 55
Graph showing resistance against length for a wire

Temperature and resistance in conductors and thermistors

Temperature *always* affects conduction, no matter whether the material is a conductor, an **insulator** or a semiconductor. In conductors the resistance increases as the temperature increases.

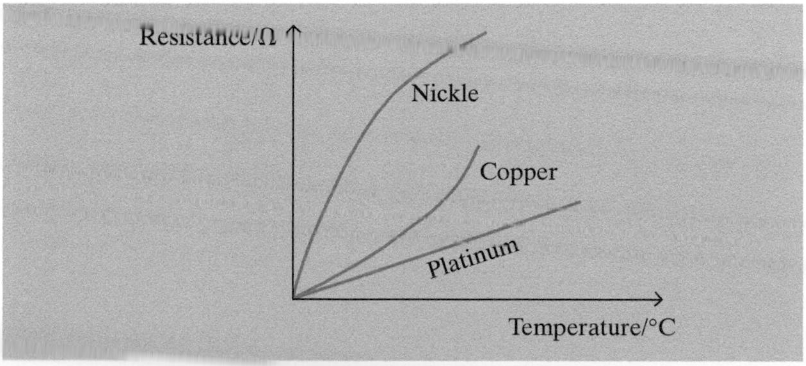

Fig 56
Graph showing the increase in resistance of a conductor with increased temperature

A metal has free electrons that move through the metal when a p.d. is applied, causing a current to flow. A metal also has vibrating positive ions. Electrons collide with these ions, causing the wire to have resistance to current.

As a metal is heated, the vibrations of the positive ions increases in **amplitude** and **frequency**. There are a more collisions between the conduction electrons and the positive ions. This increases the resistance.

Semiconductors

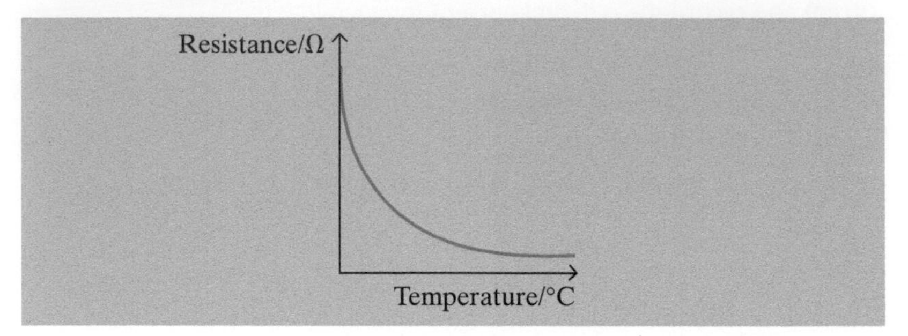

Fig 57
In thermistors, resistance decreases as temperature increases

Essential Notes

A thermistor is a device used for temperature measurement and control. The circuit symbol for a thermistor is –[/]–

Thermistors are resistors with a resistance that changes markedly with temperature. They are made from semiconducting material. A semiconductor has far fewer free electrons available for conduction than there are in a metal, and the resistivity is much therefore much higher. As the semiconductor is heated, more electrons are freed from the positive ions and the number of charge carriers (mobile electrons) therefore goes up. The resistivity of the material decreases, and hence the resistance of the thermistor falls.

Essential Notes

At higher temperatures the ions of the semiconductor vibrate more. This would normally cause the resistance to rise. However, the release of conduction electrons is the dominant effect. This also explains the shape of the graph.

Care is needed when passing current through thermistors. Currents produce heat and this decreases the resistance of the thermistor, allowing more current to flow. This further heats the thermistor, producing further resistance changes and the process can continue until the component overheats and burns out or melts.

Superconductors

If the temperature of a conductor is reduced so that it approaches absolute zero (0 K or −273 °C), the electrical resistance disappears completely. The material is said to have become a **superconductor**. Its resistivity has dropped to zero and an electric current can pass through without transferring any energy to the conductor. The temperature at which the material becomes superconducting is known as the **critical temperature**, T_c.

The critical temperatures for metal superconductors are typically close to absolute zero, 1 to 4 K. Ceramic superconductors now exist that have critical temperatures as high as 125 K (−148 °C) (Fig 58).

Superconductors have important uses, for example carrying electrical power without losses, and constructing very strong electromagnets.

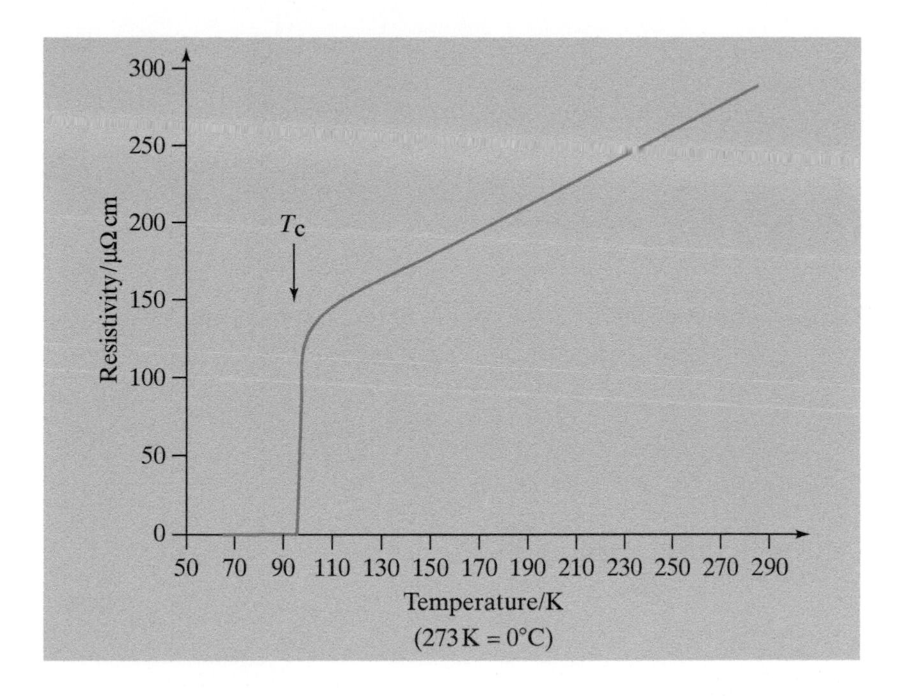

Fig 58
Resistivity against temperature for a 'high-temperature' superconductor

3.5.1.4 Circuits

This section is about the behaviour of electric circuits. The symbols for the components that you need to be familiar with are shown in Fig 59.

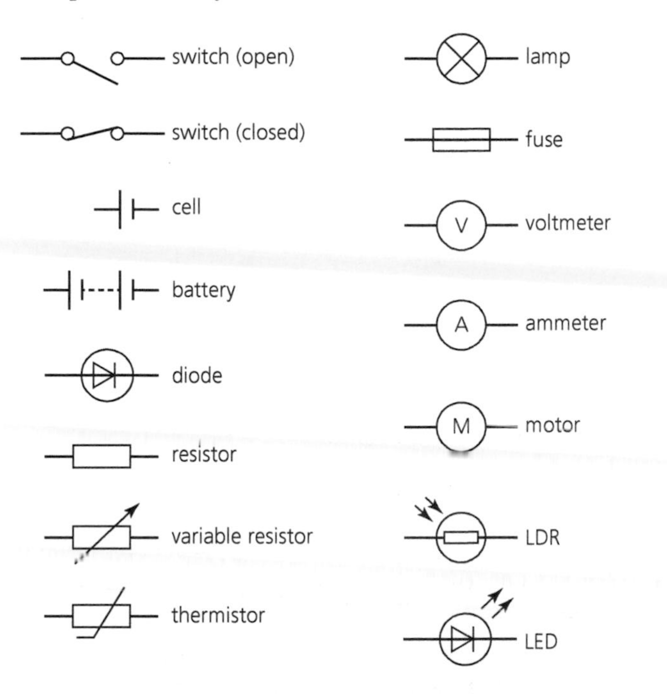

Fig 59
Circuit symbols

Resistors in series

In series, the same current (I) flows through each resistor because, by conservation of charge, the current in any part of a series circuit is the same (Fig 60, overleaf).

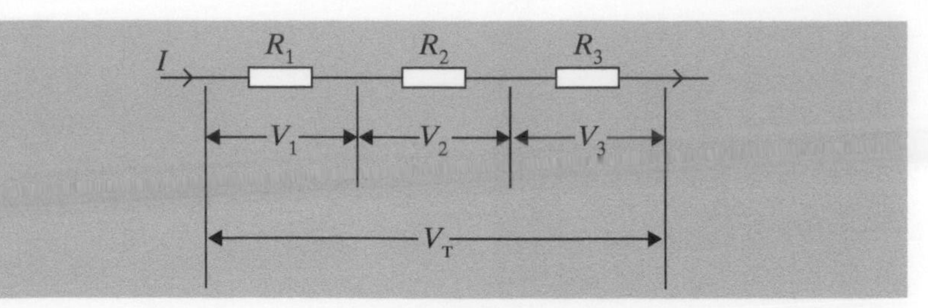

Fig 60
Three resistors in series

The p.d.s across the individual resistors add up to the total applied p.d.

$$V_T = V_1 + V_2 + V_3$$

Hence:

$$IR_T = IR_1 + IR_2 + IR_3$$

Dividing by I, the total resistance of any number of resistors is given by:

$$R_T = R_1 + R_2 + R_3 \ldots$$

Resistors in parallel

When resistors are in parallel, the p.d. across each resistor is the same (Fig 61).

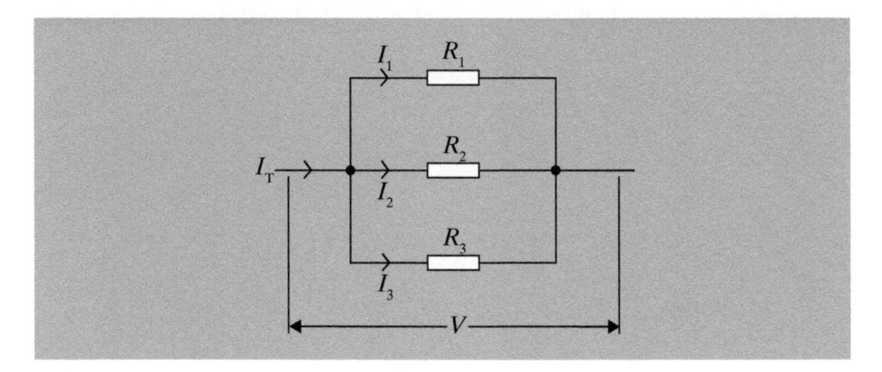

Fig 61
Three resistors in parallel

The total current is the sum of the currents through each resistor:

$$I_T = I_1 + I_2 + I_3$$

Since the potential difference is the same across all components that are connected in parallel:

$$\frac{V}{R_T} = \frac{V}{R_1} + \frac{V}{R_2} + \frac{V}{R_3}$$

Dividing by V, the total resistance of any number of resistors is given by:

$$\frac{1}{R_T} = \frac{1}{R_1} + \frac{1}{R_2} + \frac{1}{R_3}$$

Example

Draw diagrams to illustrate how three 10 Ω resistors can be connected in four different ways. Calculate the total resistance of each network of resistors.

Answer

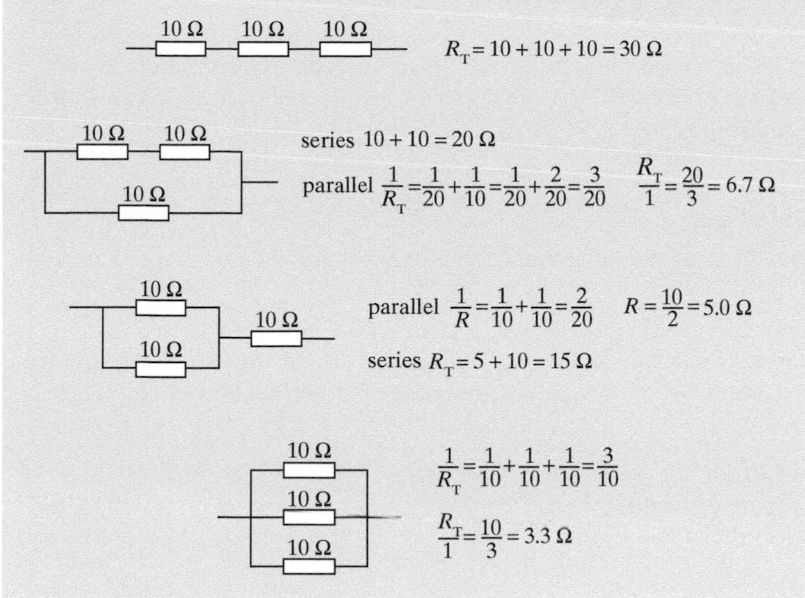

Cells in series and in parallel

Cells can be joined together to form a battery in two ways:

- **In series.** Cells are connected from the positive terminal of one cell, to the negative terminal of the next. This arrangement increases the potential difference across the terminals of the battery. The total p.d. of the battery can be found by adding the p.d. across each cell; $V_{Total} = V_1 + V_2 + V_3$. (The internal resistance of the battery is found by adding the internal resistances of each cell $R_{Total} = R_1 + R_2 + R_3 + R_4 + ...$ etc.; there is more about internal resistance in page 73)
- **In parallel.** Cells are connected together so that all the positive terminals are connected to a single point and all the negative cells are connected together to a different point. This arrangement means that the p.d. across the battery is the same as the p.d. across a single cell (if the p.d. across each individual cell is the same). The benefit of this arrangement is to increase the capacity of the battery, i.e. it will last longer!

Conservation of charge and energy in circuits

In all circuits, electric charge is conserved, i.e. all the charge which arrives at a point must leave it. Current is a flow of charge, so this can be stated as follows.

Definition

At any point in a circuit where conductors join, the total current towards the point must equal the total current flowing away from the point.

or

The algebraic sum of currents at a junction is zero (Fig 62, overleaf).

Essential Notes

This statement is known as Kirchhoff's First Law.

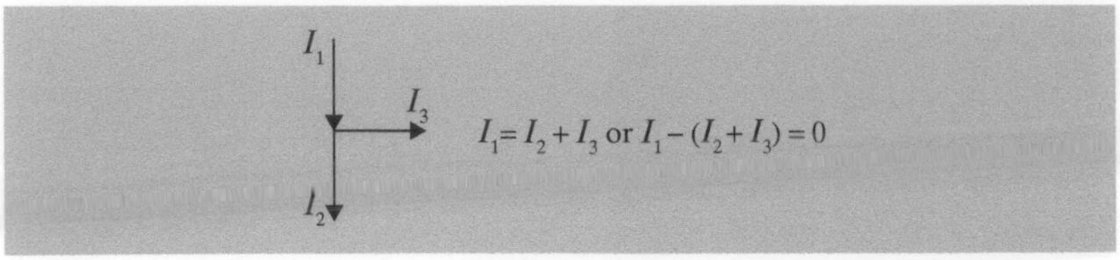

Fig 62
The sum of the currents at a junction is zero.

$I_1 = I_2 + I_3$ or $I_1 - (I_2 + I_3) = 0$

In circuits, energy differences are expressed as potential differences and measured in volts. Energy is always conserved in all circuits. This results in the following rule.

Essential Notes

This statement is known as Kirchhoff's Second Law.

Definition

The algebraic sum of potential differences around a closed circuit is zero.

A closed circuit can contain a cell or battery – in this case the electromotive force (see p. 74) must equal the sum of the p.d.s around the circuit (Fig 63).

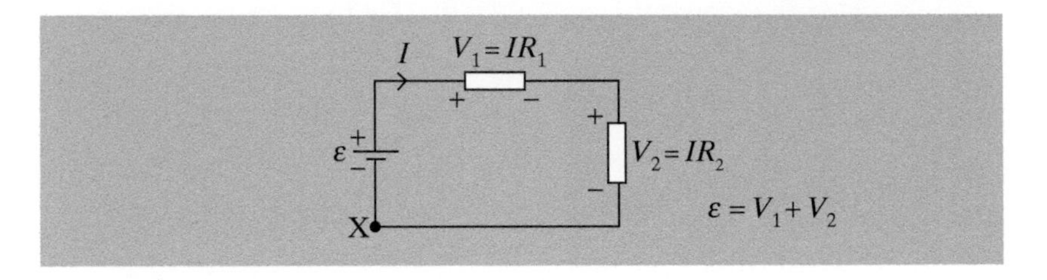

Fig 63
The electromotive force is the sum of the potential differences around the circuit.

The algebraic sum of p.d.s around a closed circuit is zero. Therefore, going round the circuit in Fig 63 in a clockwise direction (with the current) starting and finishing at point X:

$$\varepsilon - V_1 - V_2 = 0$$

or:

$$\varepsilon = V_1 + V_2$$

If there are no electromotive forces in the identified closed loop then the sum of the p.d.s across individual components must equal zero (Fig 64).

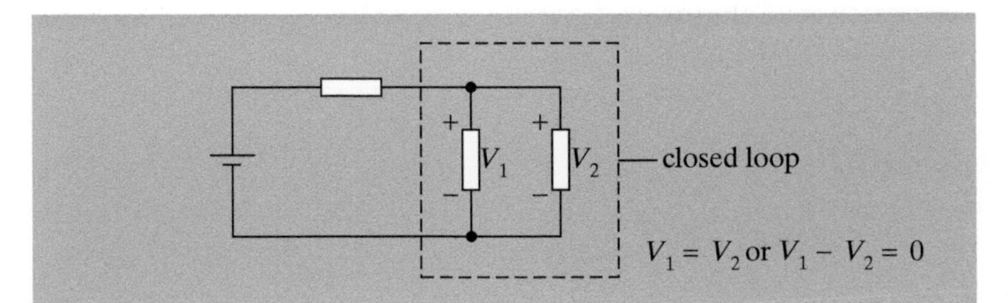

Fig 64
When there are no electromotive forces in a closed circuit, the sum of the p.d.s across the individual components is zero.

Example

Two 3 V, 0.3 A lamps are connected in parallel. Both lamps are to be run at their normal voltages from a 9 V supply by connecting a resistor in series to avoid damaging the lamps.

(a) Draw the diagram of this circuit.

(b) Calculate the current through the resistor.

(c) Calculate the voltage across the resistor.

(d) Calculate the value of the series resistor.

(e) The power rating of the resistor is 2.5 W. What is the maximum current allowed through the resistor?

(f) Comment on the suitability of using this resistor in this circuit.

Answer

(a)

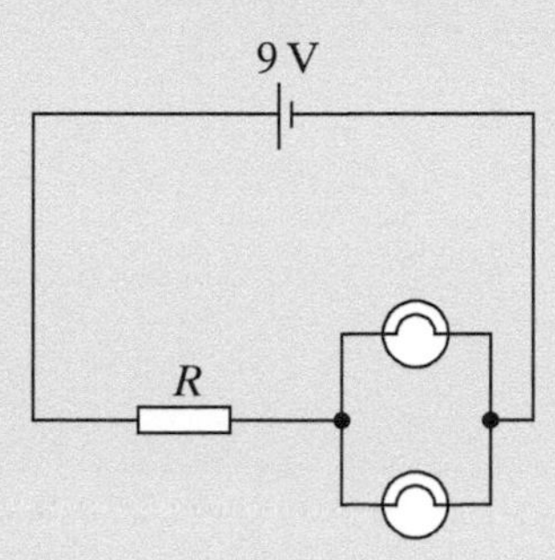

Fig 65

(b) Current through resistor = sum of currents through each lamp
$= 0.3 + 0.3 = 0.6$ A

(c) Voltage across resistor $= 9 - 3 = 6$ V.

(d) $R = \frac{V}{I} = \frac{6}{0.6} = 10\ \Omega$

(e) Rearrange $P = I^2R$ to give maximum current:

$I^2 = \frac{P}{R} = \frac{2.5}{10} = 0.25\ \text{A}^2$

$I = \sqrt{0.25} = 0.5\ \text{A}$

(f) Current which flows will generate power in excess of the power rating of the resistor. Heat generated will 'burn out' or damage the resistor.

Energy and power in d.c. circuits

To make current flow, a p.d. must exist. The p.d. is the amount of electrical energy that must be transferred to the charge and is measured in joules per coulomb, or volts.

The charge releases the gained energy as it passes through components in a circuit (e.g. lamp, motor, resistor, etc.). All the potential energy lost by the charge is ultimately changed into heat.

Energy (W) is measured in joules.

Since

$$V = \frac{W}{Q} \text{ and } Q = It$$

then the energy converted to heat is given by:

$$\text{energy change (work done) } W = VIt$$

Power (P) is the rate of change of energy, and is measured in joules per second (J s^{-1}) or watts (W):

$$\text{power } P = VI$$

Thus, the energy delivered per second (power) by a 12 V battery supplying 2 A to a circuit is 24 J s^{-1} or 24 W.

Example

Calculate the current supplied by a 1.5 V calculator battery with a power rating of 0.1 mW.

Answer

Rearrange $P = VI$ to $I = \frac{P}{V} = \frac{0.1 \times 10^{-3}}{1.5} = 0.07$ mA.

As current flows through resistors and lamps, heat is produced. The amount of power developed can be calculated using the equation:

$$P = VI$$

By substituting $V = IR$ into $P = VI$ we can arrive at an alternative equation:

$$P = I^2R$$

By substituting $I = \frac{V}{R}$ into $P = VI$ we can arrive at an alternative equation:

$$P = \frac{V^2}{R}$$

Essential Notes

The equation $P = I^2R$ is important because it shows that the heating effect is proportional to the square of the current. Therefore doubling the current will produce four times the rate of heating.

Example

(a) The power dissipated in a resistor R carrying a current I is P. If the resistance is doubled and the current halved, what power is now dissipated?

(b) A lamp is rated at 240 V, 60 W. What is its operating current?

(c) If the lamp operates at 200 V, what power is now dissipated by the lamp if its resistance remains the same when the voltage changes?

Answer

(a) Original power is given by $P = I^2R$, but $I_1 = \left(\frac{I}{2}\right)^2$ and $R_1 = 2R$

New power is given by $P_1 = \frac{I^2}{4} \times 2R = I^2\frac{R}{2}$ or $P_1 = \frac{1}{2}P$

(b) $P = VI$

Rearranging, $I = \frac{P}{V} = \frac{60}{240} = 0.25\text{ A}$

(c) $P = \frac{V^2}{R}$

Rearrange to find resistance of lamp $R = \frac{V^2}{P} = \frac{240 \times 240}{60} = 960\ \Omega$

Power of lamp $P = \frac{V^2}{R} = \frac{200 \times 200}{960} = 42\text{ W}$

3.5.1.5 Potential divider

As its name suggests, a **potential divider** splits up the potential difference (voltage) from a source. This can be done using two or more resistors in series. In this case the p.d. across the output terminals varies according to the ratio of the resistances in series (Fig 66).

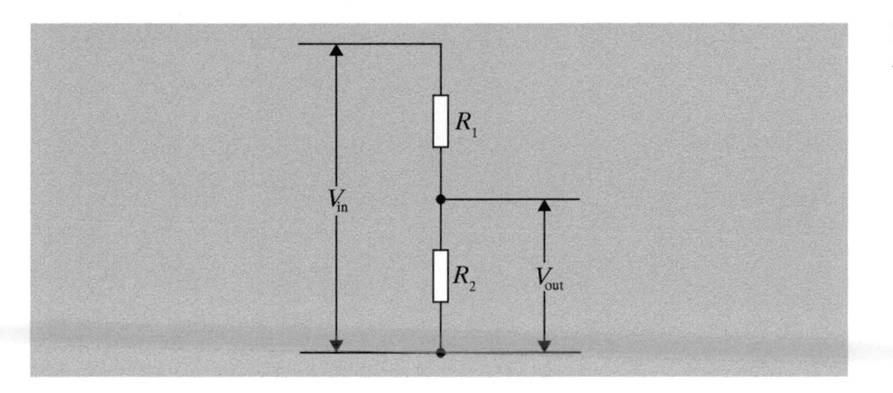

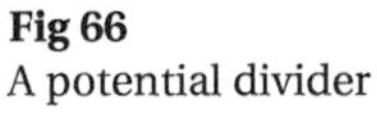
Fig 66
A potential divider

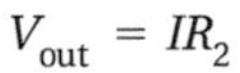

$$V_{out} = IR_2$$

$$V_{in} = I(R_1 + R_2)$$

The current I is the same in all parts of a series circuit, so:

$$\frac{V_{out}}{V_{in}} = \frac{IR_2}{I(R_1 + R_2)}$$

Therefore:

$$V_{out} = \frac{V_{in}R_2}{(R_1 + R_2)}$$

Notes

When using this equation, take care to match up the resistor values with the labelled positions of R_1 and R_2.

See Fig 67 for an example of applying the potential divider equation.

Fig 67
Calculation of V_{out} with sample values

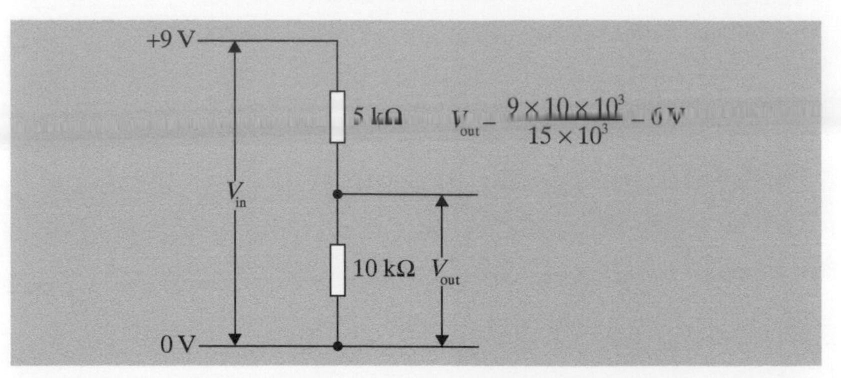

Alternatively, a slider (variable) resistance may be used (Fig 68).

Fig 68
A slider

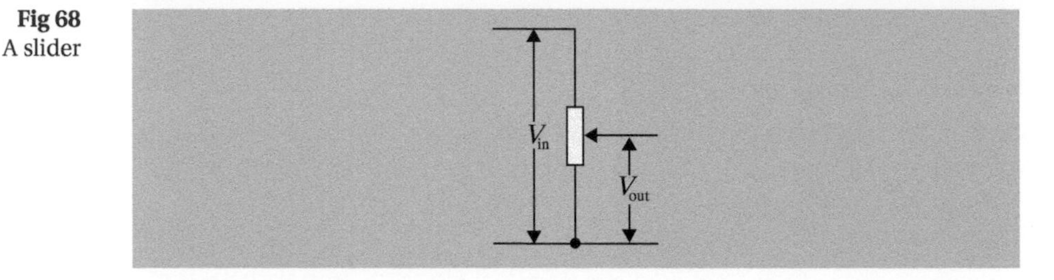

The p.d. across the output terminals varies with the position of the contact on the track (Fig 69).

Fig 69
Four examples showing variation of output voltage with contact position on a slider

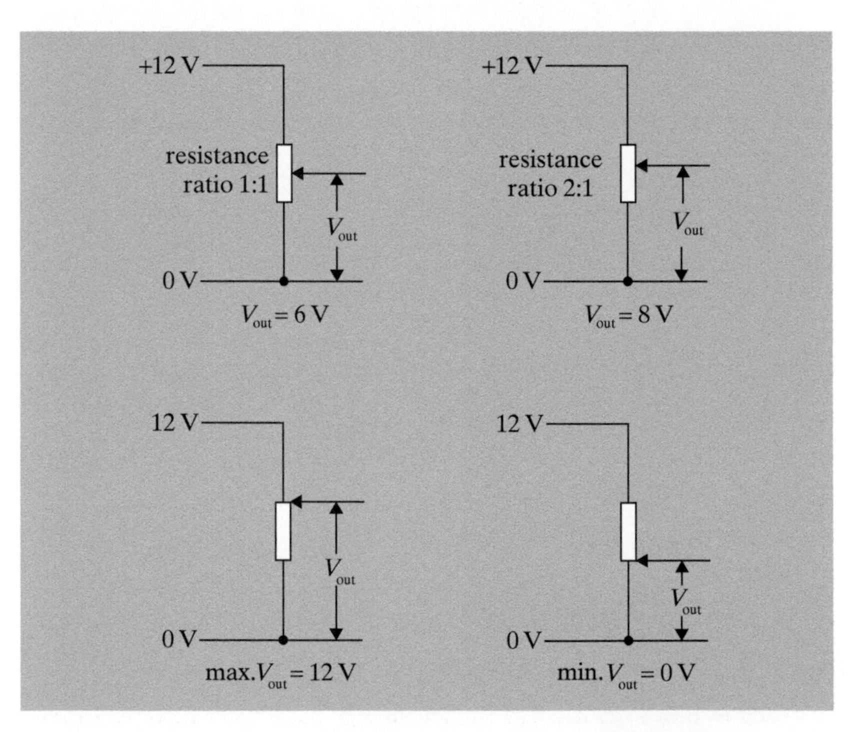

Example

(a) Calculate the potential difference between A and B in Fig 70.

(b) Calculate the potential difference between points A and B when another 5 Ω resistor is connected between them.

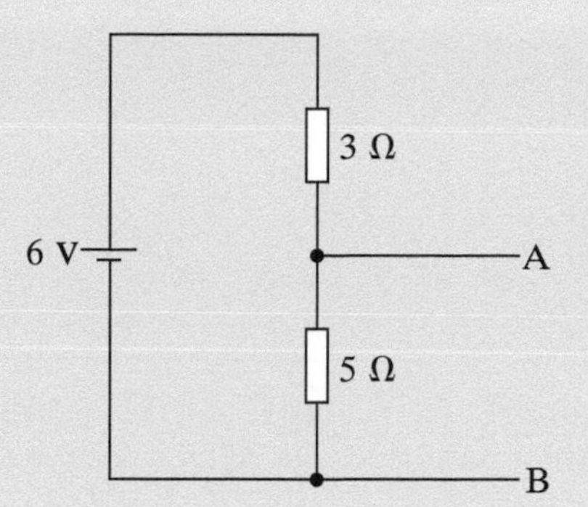

Fig 70

Answer

(a) Using the potential divider equation:

$$V_{out} = \frac{V_{in} R_2}{(R_1 + R_2)} = \frac{6 \times 5}{(3 + 5)} = 3.8 \text{ V}$$

(b)

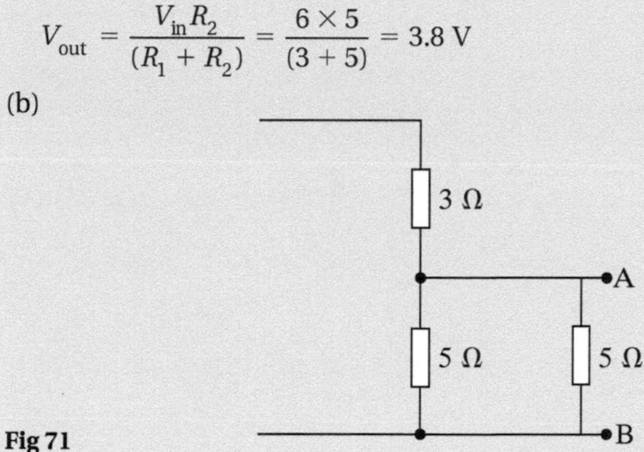

Fig 71

From Fig 71, the sum of two 5 Ω resistors in parallel is given by:

$$\frac{1}{R_T} = \frac{1}{5} + \frac{1}{5} = \frac{2}{5}$$

Therefore $R_T = \frac{5}{2} = 2.5\ \Omega$

Applying the potential divider equation:

$$V_{out} = \frac{V_{in} R_T}{(R_1 + R_T)}$$

$$V_{out} = \frac{6 \times 2.5}{(3 + 2.5)} = 2.7 \text{ V}$$

Using sensors as part of potential dividers

Light and temperature sensors can be incorporated into potential dividers to vary the output voltage depending upon ambient conditions.

A light sensor (LDR) has a low resistance in bright light and a high resistance in darkness. Using the potential divider shown in Fig 72 (overleaf), the output voltage can be made to change with light intensity.

Fig 72
A light sensor as part of a potential divider

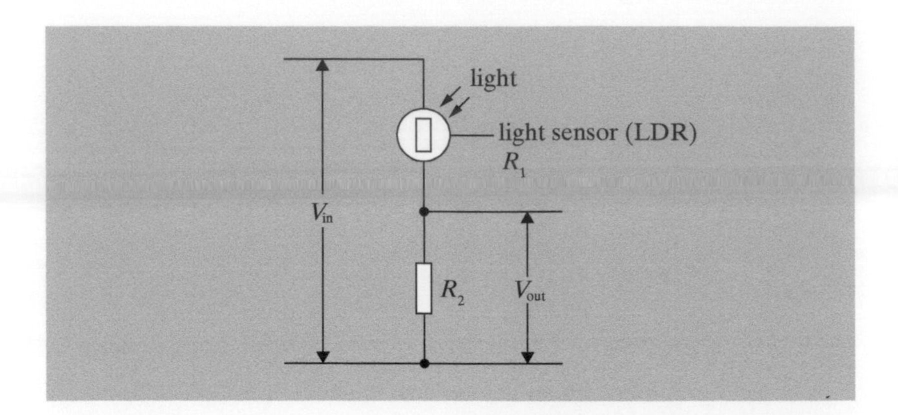

As the light gets dimmer the resistance of the LDR gets higher, the ratio of resistances changes, and so does the voltage ratio, such that the output voltage across R_2 decreases.

A temperature sensor (thermistor) has a low resistance when hot and a high resistance when cold. Using the potential divider shown in Fig 73, the output voltage can be made to change with temperature changes.

Fig 73
A temperature sensor as part of a potential divider

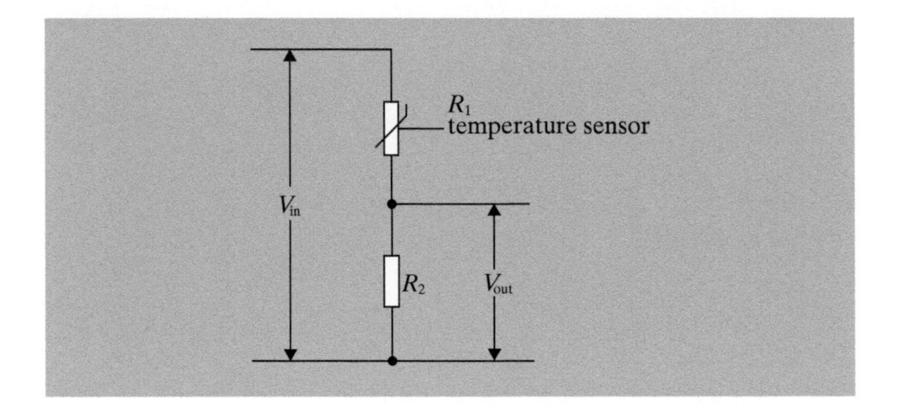

As the temperature gets colder the resistance of the thermistor increases and the voltage ratio changes, such that the output across R_2 decreases.

If the values of the resistances of the light sensor or temperature sensor are known, then the output of each potential divider under different conditions can be calculated using the equation:

$$V_{out} = \frac{V_{in} R_2}{(R_1 + R_2)}$$

Example

A potential divider is made from a light sensor (whose resistance decreases as the brightness of the light shining on it increases) and a 1000 Ω resistor (see Fig 74).

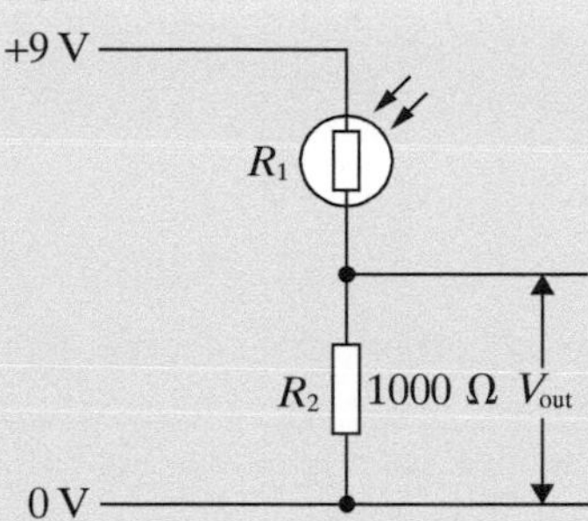

Fig 74

If the resistance of the light sensor is 500 ohms when in bright light, calculate the output voltage across R_2.

Answer

Using the equation $V_{out} = \frac{V_{in}R_2}{(R_1 + R_2)}$

$V_{in} = 9\text{ V}$, $R_1 = 500\ \Omega$ and $(R_1 + R_2) = (500 + 1000) = 1500\ \Omega$

$V_{out} = 9 \times \frac{1000}{1500} = 6\text{ V}$

Example

An LED is to be used in a portable reading lamp, designed to run from a car battery. The potential difference across the LED is 3.9 V and its operating current is 1400 mA. Specify the value and power rating of the series resistor needed to run the LED correctly, if the car battery has a terminal p.d. of 12 V and an internal resistance of 2 Ω.

Type of resistor	Power rating	Stability
Metal film	Very low at less than 3 W	High 1%
Carbon	Low at less than 5 W	Low 20%
Wire wound	High up to 500 W	High 1%

Answer

'Terminal p.d.' is simply a shortened way of saying 'potential difference across the terminals (of the battery)'. The potential differences across the components in the circuit must therefore add up to 12 V. 12 $V_R + V_{IR}$ + 3.9 V. So the p.d. across the internal resistor and the series resistor must add to 12 – 3.9 = 8.1 V. The current through the circuit is 1.4 A, The combined resistance of the two resistors = R = 8.1/1.4 = 5.79 Ω, so the series resistor has to have a value of 5.79 – 2 = 3.8 Ω. The p.d. across the series resistor is 8.1 – 2.8 = 5.3 V.

Power dissipated in the resistor = $P = IV$ = 1.4 × 5.3 = 7.4 W.

Choose a wire wound resistor.

3.5.1.6 Electromotive force and internal resistance

The **electromotive force** of a cell, or any other source of electrical energy (e.g. dynamo or thermocouple) can be defined as the p.d. across the source when no current flows, and is the energy per coulomb produced by the source. Electromotive force is usually shortened to **e.m.f.** and given the symbol ε.

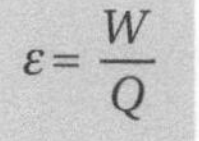

$$\varepsilon = \frac{W}{Q}$$

Fig 75
Circuit showing the internal resistance of a cell

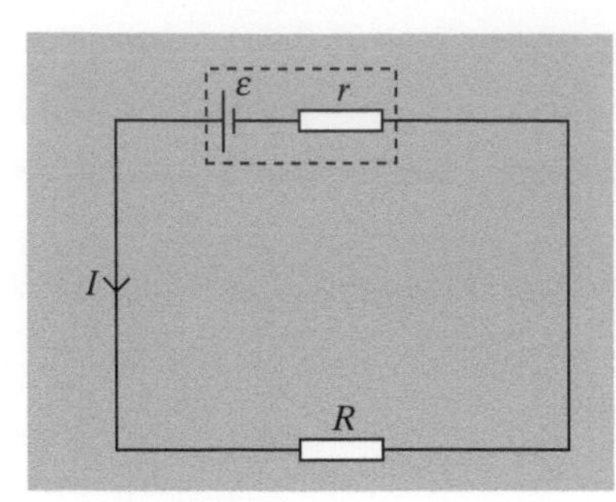

Internal resistance

The materials inside a cell, or other power source, offer a resistance to the flow of current. This is known as the **internal resistance** of the cell (usually given the symbol r) and is measured in ohms (Fig 75).

- When no current flows in the circuit then the e.m.f. = p.d. across the cell.
- When current flows in the circuit there is a p.d. across $R (V_R = IR)$ and a p.d. across $r (V_r = Ir)$.
- Both energy and charge are conserved in the circuit.
- The current is the same in any part of the circuit.
  e.m.f. = sum of the p.d.s (Kirchhoff's Second Law)

$$\varepsilon = IR + Ir = I(R + r)$$

Essential Notes

The terminal p.d. of the cell is given by

$V = IR$

and not by $V = Ir$.

Example

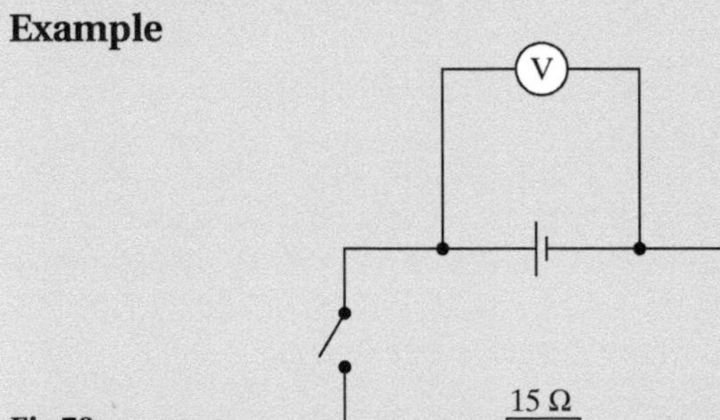

Fig 76

When the switch in the circuit in Fig 76 is open the voltmeter records 6.2 V. When the switch is closed the reading changes to 6 V.

(a) What causes the voltage across the battery to change?

(b) What is the value of the internal resistance of the battery?

(c) After the battery has been used for some time the internal resistance is 2.0 Ω. Calculate the current that flows when the switch is closed.

Answer

(a) When the switch is closed, a current, I, flows in the circuit. This current flows through the internal resistance, r, of the battery. This leads to a potential difference equal to Ir across the internal resistance, so that the voltage measured at the battery terminals is now less than when the switch was open.

(b) p.d. across internal resistance = ε − p.d. across external resistance
= 6.2 − 6 = 0.2 V

Current through circuit $I = \frac{V}{R} = \frac{6}{15} = 0.4$ A

Since it is a series circuit, there is also 0.4 A through r. So

$$r = \frac{V}{I} = \frac{0.2}{0.4} = 0.5\,\Omega$$

(c) $I = \frac{\varepsilon}{(R + r)} = \frac{6.2}{(15 + 2.0)} = 0.36\text{ A}$

Measuring the e.m.f. and internal resistance of a cell

The circuit in Fig 77 may be used to measure the e.m.f. and the internal resistance of a cell.

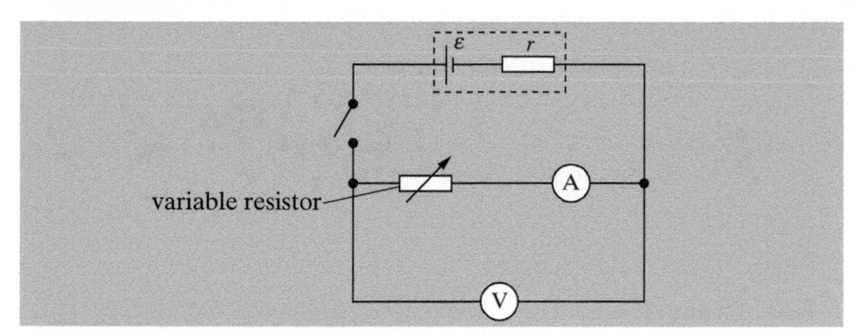

Fig 77
Circuit for measuring the e.m.f. and internal resistance of a cell

- The variable resistor is adjusted to enable a range of current (ammeter) and p.d. (voltmeter) readings to be recorded.
- A switch is used to break the circuit between readings to avoid 'running down' the cell.
- A graph is drawn (Fig 78), plotting p.d. on the y axis against current on the x axis.

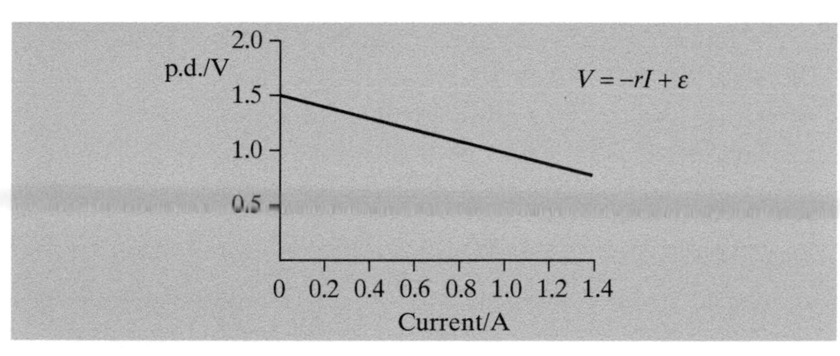

Fig 78
Graph of p.d. versus current through a variable resistor

- y axis intercept = e.m.f. (ε)
- −(gradient) = internal resistance, r

Example

Use the readings on the graph in Fig 78 to calculate:

(a) the e.m.f. of the cell

(b) the internal resistance of the cell.

Answer

(a) e.m.f. = intercept on y axis = 1.5 V

(b) internal resistance = −(gradient) = $\frac{1.5 - 0.7}{1.4} = 0.57\,\Omega$

Practical and mathematical skills

In both the AS and A-level papers at least 15% of marks will be allocated to the assessment of skills related to practical physics. A minimum of 40% of the marks will be allocated to assessing mathematical skills at level 2 and above. These practical and mathematical skills are likely to overlap to some extent, for example applying mathematical concepts to analysing given data and in plotting and interpretation of graphs.

The required practical activities assessed at AS are:

- Investigation into the variation of the frequency of stationary waves on a string with length, tension and mass per unit length of the string.
- Investigation of interference effects to include the Young's slit experiment and interference by a diffraction grating.
- Determination of g by a free-fall method.
- Determination of the Young modulus by a simple method.
- Determination of resistivity of a wire using a micrometer, ammeter and voltmeter.
- Investigation of the e.m.f. and internal resistance of electric cells and batteries by measuring the variation of the terminal p.d. of the cell with current in it.

The additional required practical activities assessed only at A-level are:

- Investigation into simple harmonic motion using a mass–spring system and a simple pendulum.
- Investigation of Boyle's (constant temperature) Law and Charles's (constant pressure) Law for a gas.
- Investigation of the charge and discharge of capacitors. Analysis techniques should include log–linear plotting leading to a determination of the time constant RC.
- Investigate how the force on a wire varies with flux density, current and length of wire using a top pan balance.
- Investigate, using a search coil and oscilloscope, the effect on magnetic flux linkage of varying the angle between a search coil and magnetic field direction.
- Investigation of the inverse-square law for gamma radiation.

Questions will assess the ability to understand in detail how to ensure that the use of instruments, equipment and techniques leads to results that are as accurate as possible. The list of apparatus and techniques is given in the specification.

Exam questions may require problem solving and application of scientific knowledge in practical contexts, including novel contexts.

Exam questions may also ask for critical comments on a given experimental method, conclusions from given observations or require the presentation of data in appropriate ways such as in tables or graphs. It will also be necessary to express numerical results to an appropriate precision with reference to uncertainties and errors, for example in thermometer readings.

The mathematical skills assessed are given in the specification.

Data and formulae

FUNDAMENTAL CONSTANTS AND VALUES

Quantity	Symbol	Value	Units
speed of light in vacuo	c	3.00×10^{8}	$\mathrm{m\,s^{-1}}$
permeability of free space	μ_0	$4\pi \times 10^{-7}$	$\mathrm{H\,m^{-1}}$
permittivity of free space	ε_0	8.85×10^{-12}	$\mathrm{F\,m^{-1}}$
magnitude of the charge of electron	e	1.60×10^{-19}	C
the Planck constant	h	6.63×10^{-34}	J s
gravitational constant	G	6.67×10^{-11}	$\mathrm{N\,m^2\,kg^{-2}}$
the Avogadro constant	N_A	6.02×10^{23}	$\mathrm{mol^{-1}}$
molar gas constant	R	8.31	$\mathrm{J\,K^{-1}\,mol^{-1}}$
the Boltzmann constant	k	1.38×10^{-23}	$\mathrm{J\,K^{-1}}$
the Stefan constant	σ	5.67×10^{-8}	$\mathrm{W\,m^{-2}\,K^{-4}}$
the Wien constant	α	2.90×10^{-3}	m K
electron rest mass (equivalent to 5.5×10^{-4} u)	m_e	9.11×10^{-31}	kg
electron charge–mass ratio	e/m_e	1.76×10^{11}	$\mathrm{C\,kg^{-1}}$
proton rest mass (equivalent to 1.00728 u)	m_p	1.673×10^{-27}	kg
proton charge–mass ratio	e/m_p	9.58×10^{7}	$\mathrm{C\,kg^{-1}}$
neturon rest mass (equivalent to 1.00867 u)	m_n	1.675×10^{-27}	kg
gravitational field strength	g	9.81	$\mathrm{N\,kg^{-1}}$
acceleration due to gravity	g	9.81	$\mathrm{m\,s^{-2}}$
atomic mass unit (1 u is equivalent to 931.5 MeV)	u	1.661×10^{-27}	kg

ASTRONOMICAL DATA

Body	Mass/kg	Mean radius/m
Sun	1.99×10^{30}	6.96×10^{8}
Earth	5.97×10^{24}	6.37×10^{6}

GEOMETRICAL EQUATIONS

arc length = $r\theta$

circumference of circle = $2\pi r$

area of circle = πr^2

curved surface area of cylinder = $2\pi rh$

volume of cylinder = $\pi r^2 h$

area of sphere = $4\pi r^2$

volume of sphere = $\frac{4}{3}\pi r^3$

ALGEBRAIC EQUATIONS

quadratic equation

$$x = \frac{\left(-b \pm \sqrt{\left(b^2 - 4ac\right)}\right)}{2a}$$

PARTICLE PHYSICS

Rest energy values

Class	Name	Symbol	Rest energy /MeV
photon	photon	γ	$= 0$
lepton	neutrino	ν_e	< 0.3 eV
		ν_μ	< 0.3 eV
	electron	$e^\pm$	0.510 999
	muon	$\mu^\pm$	105.659
mesons	π meson (pion)	$\pi^\pm$	139.576
		π^0	134.972
	K meson (kaon)	$K^\pm$	493.821
		K^0	497.762
baryons	proton	p	938.257
	neutron	n	939.551

Properties of quarks (antiquarks have opposite signs)

Type	Charge relative to electron charge, *e*	Baryon number	Strangeness
u	$+\frac{2}{3}$	$+\frac{1}{3}$	0
d	$-\frac{1}{3}$	$+\frac{1}{3}$	0
s	$-\frac{1}{3}$	$+\frac{1}{3}$	-1

Properties of leptons

	lepton number
particles: $e^-, \nu_e; \mu^-, \nu_\mu$	$+1$
antiparticles: $e^+, \overline{\nu}_e; \mu^+, \overline{\nu}_\mu$	-1

Photons and energy levels

photon energy $E = hf = hc/\lambda$

photoelectricity $hf = \phi + E_{k(max)}$

energy levels $hf = E_1 - E_2$

de Broglie wavelength $\lambda = \frac{h}{p} = \frac{h}{mv}$

ELECTRICITY

current and p.d. $I = \frac{\Delta Q}{\Delta t}$ $V = \frac{W}{Q}$ $R = \frac{V}{I}$

e.m.f. $\varepsilon = \frac{E}{Q}$ $\varepsilon = I(R + r)$

resistors in series $R = R_1 + R_2 + R_3 + \ldots$

resistors in parallel $\frac{1}{R} = \frac{1}{R_1} + \frac{1}{R_2} + \frac{1}{R_3} + \ldots$

resistivity $\rho = \frac{RA}{L}$

power $P = VI = I^2R = \frac{V^2}{R}$

alternating current $I_{rms} = \frac{I_0}{\sqrt{2}}$ $V_{rms} = \frac{V_0}{\sqrt{2}}$

MECHANICS

moments: moment $= Fd$

velocity and acceleration: $v = \dfrac{\Delta s}{\Delta t}$ $\quad a = \dfrac{\Delta v}{\Delta t}$

equations of motion: $v = u + at$ $\quad s = \dfrac{(u+v)}{2}t$

$v^2 = u^2 + 2as$ $\quad s = ut + \dfrac{at^2}{2}$

force: $F = ma$

work, energy and power: $W = Fs\cos\theta$

$E_k = \frac{1}{2}mv^2$ $\quad \Delta E_p = mg\Delta h$

$P = \dfrac{\Delta W}{\Delta t}$ $\quad P = Fv$

$$\text{efficiency} = \frac{\text{useful output power}}{\text{input power}}$$

MATERIALS

density: $\rho = \dfrac{m}{V}$ $\quad$ Hooke's Law $F = k\Delta L$

$$\text{Young modulus} = \frac{\text{tensile stress}}{\text{tensile strain}}$$

$$\text{tensile stress} = \frac{F}{A}$$

$$\text{tensile strain} = \frac{\Delta L}{L}$$

energy stored: $E = \frac{1}{2}F\Delta L$

WAVES

wave speed: $c = f\lambda$ $\quad$ period $T = \dfrac{1}{f}$

first harmonic for a standing wave on a string $f = \dfrac{1}{2l}\sqrt{\dfrac{T}{\mu}}$

fringe spacing: $w = \dfrac{\lambda D}{s}$ $\quad$ diffraction grating $d\sin\theta = n\lambda$

refractive index of a substance s: $n = \dfrac{c}{c_s}$

for two different substances of refractive indices n_1 and n_2,

law of refraction: $n_1\sin\theta_1 = n_2\sin\theta_2$

critical angle: $\sin\theta_c = \dfrac{n_2}{n_1}$ for $n_1 > n_2$

Practice exam-style questions

Multiple-choice questions

Select the best answer from the 4 alternatives.

1 An electric kettle, which is designed to run in the UK at a domestic voltage of 230 V, has a power rating of 2.2 kW. Which of the following statements is not true?

A The electrical resistance of the kettle is approximately 24 Ω.

B The electric current flowing through the kettle is approximately 9.6 A.

C In 5 minutes the kettle will transfer 0.18 kWh of energy.

D In 5 minutes the kettle will transfer 11 kJ of energy.

2 The graphs show how the resistance of different materials changes with temperature.

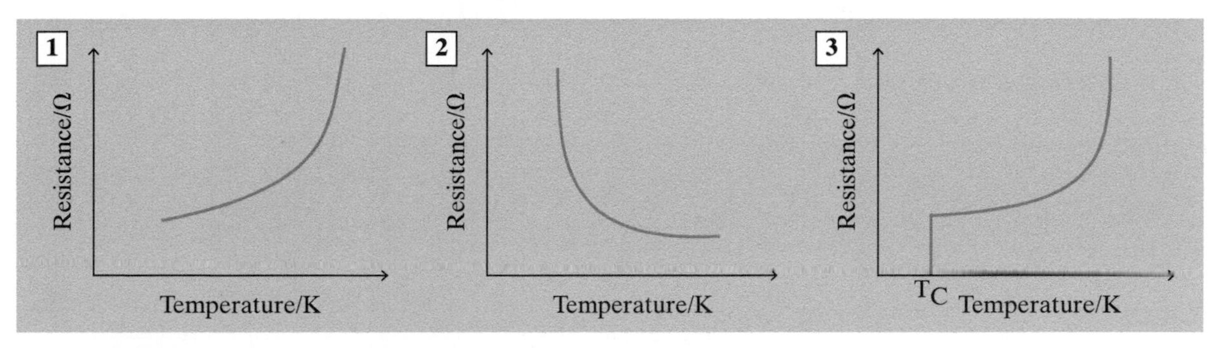

	Graph 1	Graph 2	Graph 3
A	Metal	Superconductor	Semi-conductor
B	Metal	Semi-conductor	Superconductor
C	Semi-conductor	Metal	Superconductor
D	Superconductor	Semi-conductor	Metal

3 A small spring, of spring constant k, is stretched by an extension x, in a toy to fire a small mass, m, vertically up in the air. Which of the following changes will cause the largest increase in the height reached by the mass?

A Use a spring with spring constant $2k$.

B Double the extension, to $2x$.

C Use a smaller mass, ½ m.

D Use a larger mass, $2m$.

4 A small electric motor of power 3 W is designed to run at a potential difference of 12 V. Unfortunately the only power supply available has an e.m.f. of 24 V, and an internal resistance of 3 Ω. In which of these circuits would the motor run at its correct rating?

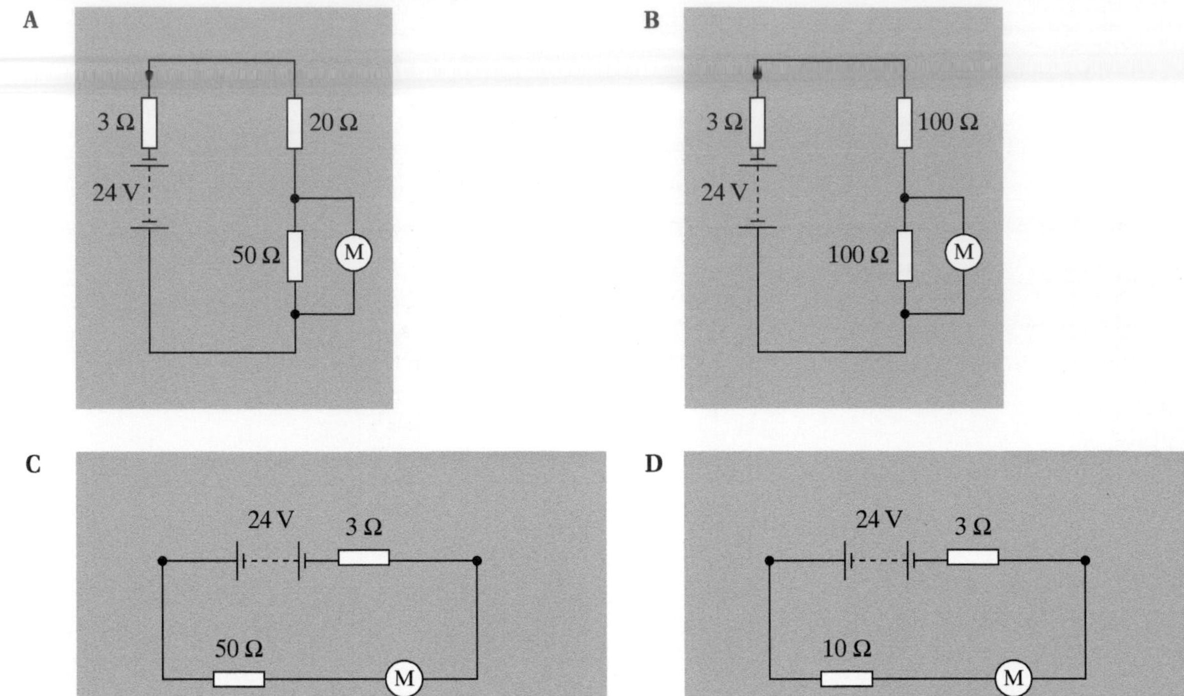

5 (a) (i) The *moment* of a force is its turning effect about a given point. Explain how the *moment* is calculated.

____

____ 2 marks

(ii) Explain what is meant by the *principle of moments*.

____

____ 1 mark

(iii) Use the principle of moments to determine the maximum load that can be lifted by the crane shown below. You may ignore the weight of the crane. Give your answer in newtons.

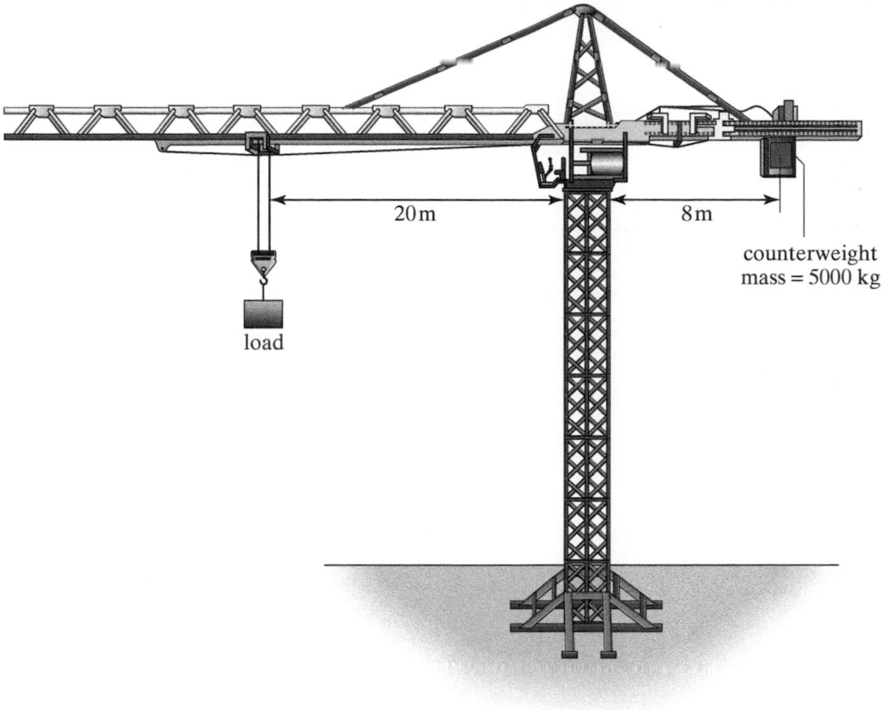

20 m

8 m

counterweight
mass = 5000 kg

load

____

____

____ 3 marks

(b) State one adjustment that could be made to the crane to allow it to lift a heavier load.

____ 1 mark

Total Marks: 7

6 A bungee jumper jumps from a platform. His feet are tied to a strong rubber cord of length 30 m. The mass of the bungee jumper is 80 kg.

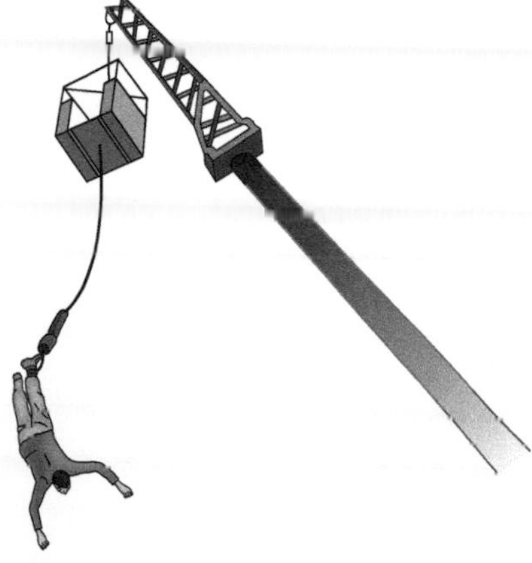

(a) **(i)** Calculate the gravitational potential energy lost by the bungee jumper as he falls between the platform and the point when the cord reaches its natural (unstretched) length.

____

____ 2 marks

(ii) What is the jumper's kinetic energy at that point?

____ 1 mark

(iii) What assumption have you made in your answer to (ii)? Explain why this assumption is justified.

____

____ 2 marks

(b) The bungee cord stretches until it is twice its natural length and brings the jumper to rest.

(i) By considering your answers to parts (a)(i) and (ii), calculate how much energy is now stored as elastic strain energy in the cord.

____ 1 mark

(ii) Calculate the tension in the cord at this point.

____

____ 2 marks

(iii) Explain why this tension is greater than the jumper's weight.

____

____ 1 mark

Total Marks: 9

7 In an attempt to measure the acceleration due to gravity, *g*, Steve is dropping stones off a vertical cliff. He starts his stopwatch as he drops a stone, and stops it when he hears it hit the beach below. He gets the following readings:

2.99
3.12
3.48
2.96
3.34

(a) What value of *g* would Steve get from these readings?

____

____

____

____ 4 mark

(b) Steve is doubtful about the accuracy and precision of his results. He wonders whether the mass or area of the stones will affect either of these.

Answer Steve's queries and discuss the main random and systematic errors which are likely to occur in his method and estimate the accuracy and precision of his results.

6 marks

Total Marks: 10

8

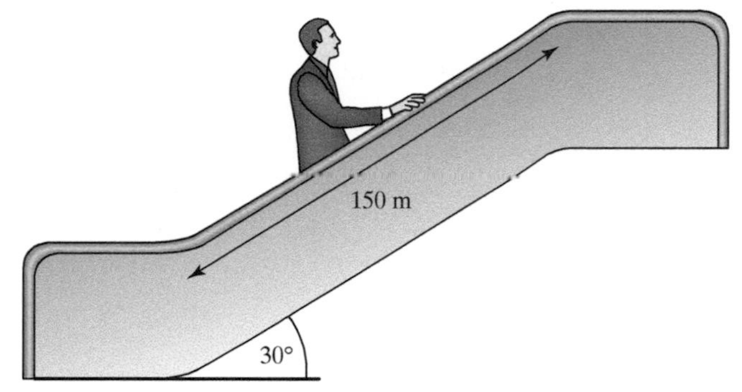

An escalator travels at 0.5 m s^{-1} and rises at an angle of 30° to the horizontal. The escalator is 150 m long.

(a) A man of mass 80 kg travels on the escalator. How much work is done by the escalator in raising the man from the bottom to the top of the escalator?

2 marks

(b) **(i)** The escalator can lift a maximum total mass of 10 000 kg at this speed. What is its useful maximum output power?

2 marks

(ii) The input power to the escalator is 45 kW. What is the efficiency of the escalator when it is lifting its maximum load?

1 mark

(c) Suggest why the efficiency of the escalator is less than 100%.

____

____________________ 2 marks

(d) The average efficiency of the escalator, calculated over the course of a day, is much less than the answer to (b)(ii). Discuss why this might be.

____

____

____________________ 2 marks

Total Marks: 9

9 An incandescent electric light bulb has a thin metal wire, known as a filament, which glows white hot when a current passes through it. When it is connected to a potential difference of 230 V and it has reached its working temperature, the electric current through the filament is 0.50 A.

(a) Explain what is meant by *electric current*.

____

____________________ 2 marks

(b) Potential difference is measured in volts. Give a definition of the volt.

____

____________________ 2 marks

(c) Calculate the resistance of the light bulb filament at its working temperature.

____

____

____________________ 3 marks

(d) The light bulb is switched on all day, for 24 hours. How much electrical energy would it transfer (to heat and light) in this time?

____

____

____

____________________ 4 marks

(e) For a short time, when the bulb is first switched on, the current is much higher than 0.5 A. Suggest why this is.

3 marks

Total Marks: 14

10 A student is asked to find the resistivity, ρ, of nichrome, which is a metal alloy. The student is given a 2 m length of nichrome wire which has a diameter of approximately 0.5 mm.

Describe how the student could find the resistivity. Explain what measurements need to be taken and what measuring instruments would be used. Include a suitable circuit diagram. Explain what steps the student could take to get an answer which is accurate and reliable.

(You will be assessed on your ability to communicate your answer clearly.)

Total Marks: 7

11 During a game of cricket a batsman hits the ball for six runs. The ball travels through the air in a parabolic path and lands 80 m away from the batsman.

(a) The ball leaves the bat with a horizontal velocity of 40 m s^{-1}. How long is the ball in the air?

2 marks

(b) What is the greatest height reached by the ball?

3 marks

(c) How far from the batsmen could a fielder stand so as to catch the ball when it is 2 m above the ground, (there are two possible positions).

4 marks

(d) What assumptions have you made in order to simplify the calculation? Are these likely to be valid?

4 marks

Total Marks: 13

12 In an experiment to measure the Young modulus, E, of copper, two identical pieces of copper wire, A and B, are suspended vertically from the same support and an identical mass M kg is hung from the end of each.

Extra masses are then hung on the end of wire A and its extension is measured by comparing it to the length of wire A, using a vernier scale with a resolution of 0.01 mm.

The diameter of the wire was measured with a micrometer and the following results were obtained.

0.68 mm	0.74 mm	0.68 mm	0.68 mm	0.67 mm	0.67 mm

(a) The experiment uses **two** lengths of wire, rather than just one, to reduce two potential sources of systematic error. Explain this.

2 marks

(b) Why is the diameter measured a number of times?

2 marks

(c) Which measurement in the experiment is likely to have the greatest uncertainty? Discuss what you could do to improve the precision in this reading. What are the possible disadvantages with your suggestion?

4 marks

(d) State and explain two safety precautions you would take when carrying out this experiment.

2 marks

(e) Calculate the cross-sectional area of the wire and estimate the percentage uncertainty in your value.

3 marks

(f) The measurements were used to calculate series of stress and strain values for the wire. A graph of stress and strain was plotted for the wire.

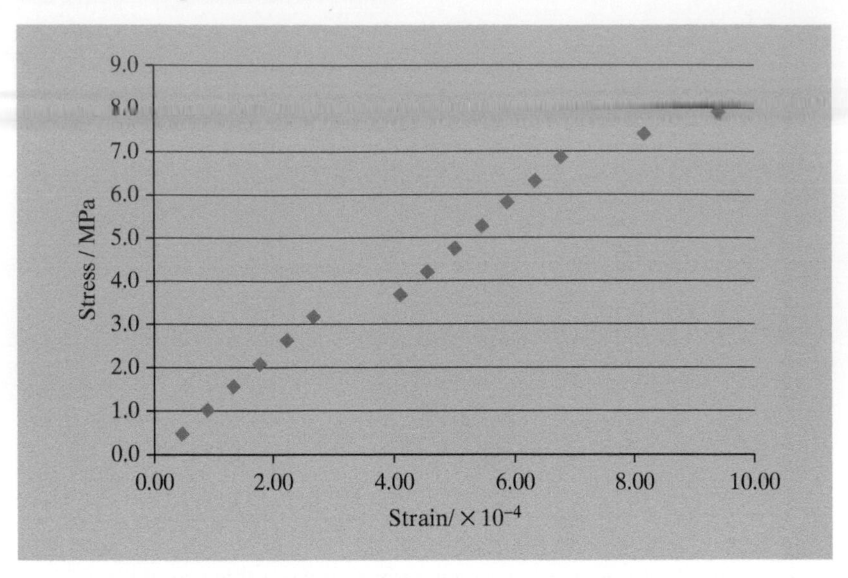

(i) Define tensile stress and state the S.I. unit which is used to measure it.

______________________________ 1 mark

(ii) Define tensile strain and state the S.I. unit which is used to measure it.

______________________________ 1 mark

(iii) Explain why stress and strain are used, rather than simply force and extension.

____

____

______________________________ 2 marks

(iv) Describe and explain the shape of the graph in as much detail as possible.

____

____

____

______________________________ 4 marks

(v) Use the graph to calculate the Young modulus and the yield stress.

____

____

______________________________ 4 marks

Total Marks: 25

Answers

Question	Answer	Marks
1	A $P = V^2 / r$ so $r = 230^2 / 2200 \approx 24\ \Omega$. B $I = P / V = 2200 / 230 \approx 9.6$ A. C Energy (kWh) = power (kW) × time (hours) = 2.2 × (5 / 60) = 11 / 60 = 0.18 kWh. D Energy (J) = power (W) × time (seconds) = 2200 × 5 × 60 = 660 kJ D is false. Therefore D is the right answer.	1
2	B	1
3	The energy stored in a stretched spring equals ½ fx = ½ kx^2. This is transferred to gravitational potential energy = mgh, so assuming 100% efficiency, h = ½ kx^2 / mg. B Doubling the extension will increase the energy stored by a factor of 4.	1
4	A The resistance of the motor, R, is given by $R = V^2 / P = 144 / 3 = 48\ \Omega$. A This is in parallel with a 50 Ω resistor, so the resistance is $1 / R = 1 / 50 + 1 / 48 \approx 2 / 50$, $R = 25\ \Omega$. This is part of the potential divider, the other half of which is 20 + 3 = 23 Ω. If these are roughly equal, there is the potential of 12 V across the motor and it will operate at the rated value. B The same working as in A, shows that the two resistors in this potential divider are 100 Ω and $1 / 100 + 1 / 48 = 1 / R$, $R \approx 33\ \Omega$, so the potential difference across the motor would be around ¼ of 24 V = 6 V. C The total current in the circuit is 24 / 101 ≈ 1 / 4 = 0.25 A. The potential across the motor would be 0.25 × 24 ≈ 6 V. D As in C, but the current through the motor is 24 / 61 = 0.4 A, potential difference = 0.4 × 48 = 19.2 V.	1
5 (a)(i) **(ii)** **(iii)**	The moment of a force is calculated using: moment = magnitude of the force × perpendicular distance of the line of action of the force from the point (2) The principle of moments states that a body can only be in equilibrium if the sum of the moments about any point is zero (i.e. sum of the clockwise moments about any point is equal to the sum of the anticlockwise moments about that point). (1) Clockwise moment = 5000 × 9.8 × 8 = 392 000 N m (2) So for equilibrium, anticlockwise moment = 392 000 N m Load × 20 = 392 000 N m, so load = 392 000/20 = 19 600 N (1)	6
5 (b)	*Either* move the counterweight towards the end of the horizontal beam (away from the vertical support) *or* increase the weight (or mass) of the counterweight *or* move the load nearer to the vertical support. (1)	1
		Total 7

<table>
<tr><th>Question</th><th>Answer</th><th>Marks</th></tr>
<tr><td>6 (a)(i)

(ii)

(iii)</td><td>Use $\Delta E_p = mg\,\Delta h$ (1)
$= 80 \times 9.81 \times 30 = 23\,500$ J (1)
(lose 1 mark if more than 3 s.f. given; lose 1 mark for missing or incorrect unit)
$\Delta E_k = \Delta E_p = 23\,500$ J (1)
(must be the same answer as (a) (i), allow error carried forward)
All of the jumper's (loss in) potential energy has been transferred to kinetic energy. (1)
This is justified because air resistance will be negligible (at these low speeds/for a compact/dense object). (1)</td><td>5</td></tr>
<tr><td>6 (b)(i)

(ii)

(iii)</td><td>The change in potential energy is twice the answer to (a) (i) since he has fallen twice as far, so $\Delta E_p = 47\,000$ J (1)
Since the jumper is instantaneously at rest, all the energy has been transferred to elastic strain energy. Use $E = \frac{1}{2} F\,\Delta l$, so $F = 2E/\Delta l$ (1)
$= 2 \times 47\,000/30 = 3130$ N (to 3 s.f.) (1)
The tension is higher than the jumper's weight because the jumper is not in equilibrium (it will accelerate the jumper upwards). (1)</td><td>4</td></tr>
<tr><td></td><td></td><td>Total 9</td></tr>
<tr><td>7 (a)</td><td>Average time = 3.20 s. (1)
Taking the down direction as positive, and using $s = ut + ½\, at^2$;
Then, as $u = 0$ m s^{-1} (because the stone is dropped from rest),
$s = ½\, at^2$. Mean value of $t = 3.20$, so $a = 2 \times 50 / 3.20^2$
$= 100 / 10.2 = 9.77$ m s^{-2}. (3 for correct answer with unit)
<table><tr><td>s</td><td>50 m</td></tr><tr><td>u</td><td>0 m s^{-1}</td></tr><tr><td>v</td><td>?</td></tr><tr><td>a</td><td>g</td></tr><tr><td>t</td><td>1.15 s</td></tr></table>
Essential Notes
It is important to identify what you know, and what you need to find out. Use a suvat table.</td><td>4</td></tr>
<tr><td>7 (b)</td><td>• The acceleration due to gravity (free fall) is the same for all objects, no matter what the mass. Air resistance plays a part when the object has a large area, or a small mass, or is travelling very fast. Provided Steve's stones are of reasonable size, a mass of a few hundred grams say, his choice of stone will not affect the accuracy or precision. (2)
• Precision: Steve's stopwatch has a resolution of 0.01 seconds. However, Steve's method is nowhere near this level of precision. There is a random error in Steve's reaction starting and stopping the stopwatch, usually estimated as between ±0.1 s and ±0.2 s. The spread of Steve's results is from 2.96 to 3.48 s, which suggests a larger uncertainty of ±0.26 s. This is a percentage uncertainty of around 8%, but calculation of g depends on t^2, so the uncertainty is approximately 16%. (2)
• Accuracy: The systematic errors include the time taken for sound to travel 50 m, back to Steve. This adds approximately 50/330 = 0.15 s to each of his results, affecting the accuracy of his results.
Comparing Steve's answer to the accepted value gives an accuracy of around 5% ... he got lucky! (2)</td><td>6</td></tr>
<tr><td></td><td></td><td>Total 10</td></tr>
</table>

Question	Answer		Marks
8(a)	Work done = $Fs \cos\theta = 80 \times 9.81 \times 150 \times \cos 60°$ $= 58\,800$ J (60° is the angle between the force (weight) and the displacement *Or* work done = E_p gained by man = $mg\,\Delta h$ $= 80 \times 9.81 \times 150 \times \sin 30°$ $= 58\,800$ J	(1) (1) (1) (1)	2
8 (b)(i) (ii)	Power = $Fv \cos\theta = 10\,000 \times 9.81 \times 0.5 \cos 60°$ = 24.5 kW *or* Time to lift mass = d/v = 150/0.5 = 300 s Power = work done / time = $(10\,000 \times 9.81 \times 150 \times \cos 60°)/300$ = 24.5 kW Efficiency = useful power out / total power in = 24.5/45 = 54.4%	(1) (1) (1) (1) (1)	3
8 (c)	Energy is lost as heat due to friction/work done against friction and as heat in electrical wires. *Or* energy is needed to lift escalator itself which has significant weight (though this is balanced by an equal weight falling because the escalator is a loop).	(1) (1)	2
8 (d)	The escalator sometimes runs at less than full capacity (i.e. sometimes there are fewer people on the escalator). This reduces the useful output power. /The energy losses still occur.	 (1) (1)	2
			Total 9
9 (a)	Electric current is defined as the rate of flow of charge (or the charge that flows past a point in a given time)	(1) (1)	2
9 (b)	One volt is the potential difference between two points when one coulomb of charge transfers one joule of energy as it moves between them (1 volt = 1 joule per coulomb)	 (1) (1)	2
9 (c)	Resistance $R = V/I$ $= 230/0.5 = 460\ \Omega$ (one mark is for the correct unit, Ω)	(1) (2)	3
9 (d)	Power $P = IV = 0.5 \times 230 = 115$ W; which is 115 joules per second There are $24 \times 60 \times 60 = 86\,400$ seconds in a day so the energy transferred is $115 \times 86\,400 = 9.94 \times 10^6$ J (one mark is for the correct unit)	(1) (1) (2)	4
9 (e)	Filament / bulb is cold at first so its resistance is lower so more current flows, since $I = V/R$	(1) (1) (1)	3
			Total 14

Question	Answer	Marks
10	Circuit diagram: includes the wire, a power supply (variable), an ammeter in series with the wire and a voltmeter in parallel with it (1)	1
	Measurements: Measure the voltage and current (1) Use these to find the resistance of the wire (1) Measure the length of the wire with a metre ruler or tape (1) Measure the diameter of the wire with a micrometer (1) Use the equation $\rho = AR/l$ to find ρ (1) (at least 3 of the above)	3
	To get an accurate answer: Measure diameter several times in different places / directions and find the mean (1) Use a long piece of the wire (1) Use a low current to keep the temperature of the wire down (1) Take several sets of readings for different lengths l and find mean value of ρ (1) Plot a graph of R vs l and find ρ from gradient (1) (at least 3 of the above)	3
		Total 7
11 (a)	Although this is two-dimensional problem, we can solve it by treating the horizontal and vertical velocity separately. First identify what you know from the question.	

Vertical	Horizontal
s = ?	s = 80 m
u = ?	$u = 40\ \text{m s}^{-1}$
v = ?	v = ?
$a = -9.81\ \text{m s}^{-2}$	$a = 0\ \text{m s}^{-2}$
t = ?	t = ?

Essential Notes

The values of t are the same. There is only one time of flight.

Question	Answer	Marks
11 (a)	Time = horizontal distance / horizontal speed = 80 / 40 = 2 s.	2
11 (b)	The greatest height will be reached after half the total flight time, i.e. after 1 second, at which time the vertical velocity will be zero. Use the equation $v = u + at$ to find the initial vertical velocity, $0 = u + -9.81 \times 1$, so $u = 9.81\ \text{m s}^{-1}$. Then use $v^2 = u^2 + 2as$ to find s. $s = (v^2 - u^2) / 2a = (0 - 9.81^2) / (-9.81 \times 2) = 9.8$ m.	3
11 (c)	To find the horizontal distance when $s = 2$ m (vertically), we need to find the corresponding times. So use $s = ut + ½ at^2$. $4.9t^2 - 9.8t - 2 = 0$ Use the quadratic equation formula, with $a = 4.9$, $b = -9.8$ and $c = -2$. This gives $t = 0.23$ s or $t = 1.77$ s The horizontal speed of the ball is $40\ \text{m s}^{-1}$, so the fielder should stand at $0.23 \times 40 = 9.2$ m away (no thanks!) or $1.77 \times 40 = 71$ m away.	4

Essential Notes

This is a quadratic which in general will not factorise, so write it in the form $ax^2 + bx + c = 0$ and use the formula.

Question	Answer	Marks
11 (d)	We have assumed the ball started at ground level, which is possible but the point of impact with the bat could be as much as 2 m above the ground. We have also neglected air resistance. Although a cricket ball is small and quite dense, drag has an important part to play when the ball is moving this quickly. Air resistance would tend to reduce the maximum height and the horizontal range, and the path would not be a symmetrical parabola.	4
		Total 13
12 (a)	The extension of the wire will be quite small, so thermal expansion or a small movement of the support would be a significant error. Since a reference wire is used, any temperature change or movement of support will affect both wires and therefore will not affect the results.	2
12 (b)	The radius is measured in different places, and in different orientations, to account for any non-uniformity of the wire.	2
12 (c)	The extensions are quite small; the first measurement has uncertainty of ±4%. The extensions could be increased by using a thinner wire, but this would lead to larger uncertainties in the calculation of cross-sectional area. Increasing the tension could help but may take the copper wire past its elastic limit or even break it. Using a longer piece of wire would reduce the percentage uncertainty in the extension.	4
12 (d)	The main danger comes from the wire breaking under tension. Wearing goggles to protect eyes and using some sort of barrier to prevent people putting their feet under the weights would reduce the risk.	2
12 (e)	The average radius is 0.34 mm. (The anomalous result could be repeated.) The area of a circle is given by $\pi r^2 = 3.7 \times 10^{-7}$ m^2. Uncertainty in diameter is 0.01 / 0.67= 1.5%, so in area = 2 × 1.5% = 3%.	3
12 (f) (i)	Tensile stress is the tensile force per unit cross-sectional area. Measured in N m^{-2} or Pa. (1)	
(ii)	Tensile strain is extension per unit length. No units (a ratio of two lengths). (1)	
(iii)	Stress and strain allow comparisons to be made between materials even if the samples (wires) have different dimensions. Stress is the force per unit area, so thickness of the wire is taken into account. Strain is the extension per unit length, so the length of wire is taken into account. (2)	
(iv)	The stress versus strain graph is essentially a straight line, as you might expect for a metal. The graph shows two parallel sections. It is possible that one of the wires slipped slowly from the chuck, or a kink in the wire straightened out. The last two or three readings show extra extension. It seems likely that the wire has passed its yield point. (4)	
(v)	The Young modulus is the gradient of either straight section, e.g.: $3.18 \times 10^7 / 2.65 \times 10^{-4} = 1.2 \times 10^{11}$ Pa The yield point is where plastic deformation occurs. From the graph, this is around 70 MPa. (4) **Note** Always choose as large a section of the graph as possible when calculating the gradient.	12
		Total 25

Glossary

acceleration	the rate of change of velocity: change in velocity / time taken; unit m s^{-2}
acceleration due to gravity	the rate at which all objects accelerate under gravity if air resistance is neglected; also known as the acceleration of free-fall; on Earth it is usually taken as 9.81 m s^{-2}, but it varies slightly from place to place
accurate	when a reading is very close to the true value
air resistance	a force that acts to oppose motion through the air
amplitude	the maximum height of a wave, or the largest displacement from equilibrium
breaking stress (or ultimate tensile stress)	the maximum stress (force per unit area) that a material can withstand before it breaks
brittle	a brittle material fractures before it has undergone plastic deformation
centre of gravity	the point at which the weight of an object can be taken to act; an object will balance if it is supported at its centre of gravity
centre of mass	the point at which the mass of an object can be taken to be concentrated; in a uniform gravitational field, this is the same point as the centre of gravity
coefficient of friction	a measure of the amount of mechanical resistance that a surface exerts on an object moving across it; the magnitude of the frictional force is given by $F = \mu N$, where N is the normal reaction and μ is the coefficient of friction
component	a vector can be split into 2 or more vectors that add together to have the same effect as the original vector: these are called components; at AS/A-level this is restricted to two perpendicular components, e.g. in the horizontal and vertical directions
compression	an object in compression is under the influence of forces that tend to squash it
contact force	an electromagnetic force exerted when two solid surfaces touch each other; sometimes referred to as the 'reaction' force
conservation laws	certain physical quantities remain the same for any closed system, and these are said to be conserved quantities; during particle interactions, charge, baryon number and strangeness are conserved: the total value of each of these quantities is the same before and after any interaction
control variable	a variable that is not of direct interest in an experiment but which may have an impact on the results and so needs to be controlled (fixed)
coplanar forces	a two-dimensional system of forces that all act in the same plane; they can be drawn on a piece of paper
couple	two equal forces that act in opposite directions on an object so as to cause rotation
critical temperature	the temperature below which a material becomes superconducting
current	electric current, I, is the charge, Q, which passes a point in a given time interval, Δt: $I = \Delta Q/\Delta t$; unit A
density	the amount of mass per unit volume; symbol ρ; unit kg m^{-3}

dependent variable	the variable that changes as a consequence of a change in the independent variable in an experiment
diode	a semiconductor device that allows current to flow only in one direction
direct current (d.c.)	current produced when charges drift in a steady direction
displacement	a vector describing the difference in position of two points
drag	a resistive force, such as air resistance, which acts to oppose motion in a fluid
ductility	the ability of materials to show extended plastic deformation and become elongated under tension; a ductile metal can be drawn out into wires
efficiency	the ratio: useful energy transferred (or work done) / total energy input; this is always less than 1
elastic behaviour (elasticity)	when a material returns to its original dimensions after a deforming force is removed, it is said to behave elastically
elastic collision	a collision in which the total kinetic energy is conserved, i.e. the sum of the kinetic energies before and after the collision is equal
elastic limit	the maximum tensile force at which a material behaves elastically, i.e. returns to its original dimensions after a deforming force is removed
elastic strain energy	the potential energy stored in an elastic material that has been extended
electromotive force (e.m.f)	the energy transferred by a power source to each coulomb of charge; equal to the potential difference at the terminals of the source when no current is flowing; unit V
electron	a fundamental particle, a member of the lepton family; carries a negative charge of 1.6×10^{-19} C and has a mass of 9.11×10^{-31} kg
energy	the ability to do work, where work is defined as a force moving through a distance, for example lifting a weight: a scalar quantity, measured in joules (J)
equilibrium	an object is said to be in equilibrium if it is not accelerating
forward biased	a diode connected so that the potential difference across it allows it to conduct
free body diagram	a simplified picture of a physical situation which shows all of the relevant forces acting on a body
frequency	the number of oscillations or waves passing a point in one second, measured in hertz, Hz
friction	a force that acts between surfaces, acting so as to oppose their relative motion
gravitational potential energy	the energy stored by a mass due to its position in a gravitational field; in a uniform field, the gravitational potential energy of a mass m that is raised by a distance Δh is given by $E_p = mg\,\Delta h$
Hooke's Law	law stating that, for an object under tension, such as a wire or a spring, the extension is proportional to the applied force
hypothesis	a tentative (provisional) idea or theory to explain an observation
impulse	the magnitude of a force (F) multiplied by the time (t) for which it acts: impulse = Ft

independent variable	the variable that is deliberately altered by an experimenter
inelastic collision	a collision in which kinetic energy is not conserved
inertia	an object's resistance to acceleration; for linear motion, this is the mass
instantaneous speed	the speed (of a moving object) at a particular instant in time
instantaneous velocity	the rate of change of displacement, as measured over a very small time interval
insulator	a material with hardly any free electrons, which therefore has a very high electrical resistance
internal resistance	the intrinsic electrical resistance between the terminals of a power supply; some energy is always lost inside the supply due to this internal resistance
kinetic energy	the energy of a mass m moving at a velocity v; $E_k = \frac{1}{2}mv^2$
lift	the upward force that keeps aircraft in the air, resulting from the aircraft wings travelling at speed through the air
longitudinal wave	a wave that has oscillations parallel to the direction of travel of the wave
moment	the turning effect of a force; the moment of a force about a point is equal to: force × perpendicular distance from the line of action of the force to the point, either clockwise or anticlockwise; unit N m
momentum (*p*)	property of a moving object equal to its mass, m, multiplied by its velocity, v; $p = mv$; a vector quantity; unit kg m s^{-1}
necking	the reduction in diameter of a wire under tension when it is close to its breaking stress
newton	the S.I. unit of force; 1 newton (1 N) is the force that will accelerate a mass of 1 kg at 1 m s^{-2}
non-ohmic conductor	a material or component that conducts electricity, but with a resistance that changes as the potential difference across it is changed
nucleus	the positively charged, dense matter at the centre of every atom; formed from a combination of neutrons and protons
Ohm's Law	the current, I, through a material is proportional to the potential difference, V, across it; this holds only for certain conductors in certain conditions
ohmic conductor	a material that follows Ohm's Law, e.g. a metal at constant temperature
parallelogram law	a method for finding the resultant of two vectors
plastic behaviour	when a material is permanently deformed, even after the applied force is removed
potential difference (p.d.)	the energy transferred per unit charge (1 C) moving between two points; unit V
potential divider	an arrangement of resistors in series so that the potential difference across the combination is divided between them in the ratio of the resistors
power	the rate at which energy, E, is transferred: power $P = \Delta E / \Delta t$; sometimes expressed as work done per second; unit watt W = J s^{-1}
prediction	a forecast (from a hypothesis or theory) that can be tested by experiment
principle of conservation of energy	law stating that the total energy of a closed system is constant

principle of moments	law stating that if an object is in equilibrium, the sum of the clockwise moments about any point must equal the sum of the anticlockwise moments about that point
quadratic equation	a mathematical equation containing a variable raised to the power of 2, but no higher; for example, the equation of motion $s = ut + 1/2\ at^2$ is quadratic in the variable t (time)
resistance	a measure of how difficult it is for electric current to pass through an object; it is the ratio of potential difference, V, to current, I: resistance $R = V/I$; unit ohm, Ω
resistivity	a property of a material that describes how difficult it is for current to pass through it; for a conducting wire it is related to the resistance, R, the cross-sectional area, A, and the length, λ, by the formula: resistivity $\rho = RA/l$; unit Ω m
resolving	mathematical analysis that considers a vector as being the sum of two or more components, to facilitate calculations; typically, the resolved vector components will be at right-angles to each other
resultant	the sum of two or more vectors, such as forces
reverse biased	a diode connected so that the potential difference across it does not allow it to conduct
scalar	a physical quantity that is fully specified by its magnitude (size); it has no direction associated with it
spring constant (*k*)	the force needed to stretch a spring by unit extension: k = force/extension; unit N m^{-1}; it is a measure of stiffness of the spring
stiffness	the resistance to extension of a material under tension
strain (or tensile strain) (ε)	the fractional increase in length of a wire, l, under tension: $\varepsilon = \Delta l/l$; it has no unit
strength	a measure of the force (stress) needed to cause fracture of a material
stress (or tensile stress) (σ)	the force per unit cross-sectional area; $\sigma = F/A$
superconductor	a conductor that has zero resistance at a temperature below its critical value
tensile force	a force acting to cause extension
tensile tester	a device that records the extension of a sample of material as the load on the sample is increased
tension	an object in tension is under the influence of forces which tend to extend it
terminal velocity	the steady velocity reached by a falling object when the drag is equal to the weight
thermistor	a semiconductor device whose electrical resistance changes significantly with temperature; used as a heat sensor
torque	the rotational equivalent of force; torque produces rotational acceleration; unit N m
ultimate tensile stress (or breaking stress)	the maximum stress (force per unit area) that a material can withstand before it breaks
vector	a physical quantity that is specified by its magnitude (size) and its direction

velocity	the rate of change of displacement; velocity = change in displacement / time; unit m s^{-1}
watt	unit of power, equal to a rate of energy transfer of 1 joule per second
work	work done = force × distance moved in the direction of the force
yield point	the minimum stress at which plastic deformation occurs
Young modulus	the stiffness constant of a material, defined by the ratio: tensile stress/tensile strain

Index

acceleration 16–17
    definition 20
    due to gravity 21–23
acceleration–time graph 20
accuracy of the experiment 50
air resistance 8, 19, 24–25
    effect on projectile motion 25
algebraic equations 76
alternating current 77
amplitude 60
annihilation 37
anticlockwise moment 13
antiquarks 77
arc length 76
area
    of circle 76
    of sphere 76
    of triangle 47
1 A resistor 55
astronomical data 76
average velocity 17, 21

breaking stress 43, 46
brittle fracture 47
brittle materials 43, 47
brittleness 43
buoyancy 8

centre of gravity 15
centre of mass 15
charge carriers 54
circle
    area of 76
    circumference of 76
circuits
    algebraic sum of currents at a junction 63
    cells in series and in parallel 63
    conservation of charge and energy in 63–64
    energy and power in d.c. 65–66
    resistors in parallel 62, 77
    resistors in series 61–62, 77
    symbols 61
circumference of circle 76
clockwise moment 13
closed circuit 64
coefficient of friction 24
collisions 29–30
components of a vector 7
compression 43
conductor
    non-ohmic 56
    temperature and resistance in 59
conservation of linear momentum 29
contact forces 8
couple 14–15
    in an electric motor 15
critical angle 78
critical temperatures 60
cross-sectional area 57
current–voltage characteristics 54–57
cylinder, volume of 76

de Broglie wavelength 77
density 15, 39, 78
diffraction grating 78
direction of vector 4
displacement 16
displacement–time graphs 17–18, 20
distance 16
drag 24–25
ductile material 43, 46–47
ductility 43
dynamic friction 24

efficiency 36–37, 78
elastic collisions 30
elasticity 41, 43
elastic limit 41, 43, 46
elastic strain energy 47
electric charge 52
electric current 51–52
electromotive forces (e.m.f.) 64, 77
    of cell 72–73
electrons 51
energy 13, 35–36, 78
    conservation of 37–39
    continual transfer of 49
    stored in a spring 49
energy levels 77
energy stored 37, 47, 49, 78
equilibrium 9, 14
    alternative method of investigating 10
    conditions for 9, 13
    on a see-saw 13
explosions, conservation of momentum in 31

filament of an incandescent bulb 56
first harmonic for a standing wave on a string 78
force 32–33, 78
    at crowbar pivots 13
force–extension graphs 42
force–time graph 32
forward biased diode 56
free body diagram 9–10
frequency 60
friction 8, 24
fringe spacing 78
fundamental constants and values 76

Galileo's law of inertia 26
geometrical equations 76
gravitational potential energy 35, 38–39
Hooke's Law 41, 43
horizontal motion 23–24
hypotenuse of a triangle 7

impulse 32
inelastic collisions 30, 33–34
inertia 26
instantaneous speed 16
instantaneous velocity 17
insulator 59
internal resistance of the cell 72–73
    measuring 73
I–V characteristic 54

kilo- 53
kilowatts (kW) 35
kinetic energy 35, 37–38
Kirchhoff's second law 64, 72

law of refraction 78
leakage current 56
leptons 77
lift 25
light sensor 70
linear region 46

magnetic effect of an electric current 52
magnitude of vector 4
materials under tension 40–41
mathematical skills 74
mega- 53
metallic conductor, charges in 52
micro- 53
milli- 53
moment 78
    of a couple 14
    of a force 11–12
    principle of 12–13
momentum 26
    collisions 29–30
    conservation of linear 29
    in explosions 31
    impulse 32
    of a moving object 26
    rate of change of 27
    total linear 29
motion
    along a straight line 16–22
    equations of 78
    equations of uniformly accelerated motion 20–21
    graphs for non-uniform acceleration 19–20
    Newton's laws of 25–28
    projectile 23–25
multimeter 59

newton metres 12
newtons 8, 27

Newton's laws of motion 25–28
    first law 25–26
    second law 26–27, 32–33
    third law 28
non-ohmic conductor 56
nucleus 31

Ohmic conductors 55
Ohm's Law 57

particle physics 77
perpendicular distance 12
photoelectricity 77
photon energy 77
physical quantities 4
plasticity 43
pogo stick 49
potential differences 51–55
    algebraic sum of 64
    terminal 71
potential divider 67–71
    using sensors as part of 69–71
power 36, 77–78
    in d.c. 65–66
precision of the experiment 50
principle of conservation of energy 37
principle of moments 12–13
projectile motion 23–25
    effect of air resistance 25
proportionality limit 46

quadratic equation 23, 76
quarks 77

refractive index of a substance 78
resistance 53, 57
    internal resistance of the battery 63
    of light sensor or temperature sensor 70
    temperature and 59
resistivity 60, 77
    accurate value 59
    measuring 58–59
    unit of 57
resolving the vector 7
rest energy values 77
resultant velocity 6
rocket engine 31

scalars 4
Searle's apparatus 45
semiconductor diode 56
semiconductors 60
SI units
    of electrical quantities 53
    of resistance 53
speed 16
sphere
    area of 76
    volume of 76
spring constant 41
static friction 24
stiffness 43, 48–49
strain 42
strength 43
stress 42
stress–strain graphs 42
    of a brittle material 47
    of loading and unloading the wire 50–51
superconductors 60–61
symbols for circuits 61

temperature sensor 70
tensile forces 40
tensile strain 42
tensile stress, behaviour of materials under 43
tensile tester 42
tension 8
terminal velocity 25
therabands 40
thermistor 70
tolerance 58
total linear momentum 29

variable resistor 73
vector quantity 4
    adding 4–6
vectors 4
    components of 7
    resolution of 7
velocity 7, 16–17, 78
velocity–time graph 18–20, 25
vertical motion 23–24
voltmeter 54
volume
    of cylinder 76
    of sphere 76

water jets 33
watt 35
wave speed 78
weight of an object (in newtons) 8
work 13, 34–35, 78
work done 34–35

yield point (stress) 43, 46
Young modulus (E) 44–46, 50, 78
    experiment 45–46

zero displacement 19

Notes

Notes

Notes

Notes